THE LIBRARY
ST. MARY'S COLLEGE OF MARYLAND
ST MARY'S CITY, MARYLAND 20686

THREE MILE ISLAND:

A Reader's Guide to Selected
Government Publications and
Government-Sponsored
Research Publications

by
Peggy M. Hassler

The Scarecrow Press, Inc.
Metuchen, N.J., & London
1988

British Library Cataloguing-in-Publication data available.

Library of Congress Cataloging-in-Publication Data

Hassler, Peggy M., 1929-
   Three Mile Island : a reader's guide to selected government publications and government-sponsored research publications / by Peggy M. Hassler.
     p. cm.
   Includes index.
   ISBN 0-8108-2118-4
   1. Three Mile Island Nuclear Power Plant (Pa.)--Bibliography. 2. Nuclear power plants--Pennsylvania--Harrisburg Region--Accidents--Bibliography.   I. Title.
Z5162.A26H37   1988
[TK1345.H37]
016.36317'9--dc19   88-10086

Copyright © 1988 by Peggy M. Hassler
Manufactured in the United States of America

CONTENTS

Preface v

Introduction 1

Glossary 41

1. Pennsylvania Government and County Publications 47

2. U.S. Government Publications: Hearings, Reports and Title Lists 54

3. U.S. Government Publications: Journal Articles 103

4. U.S. Government-Sponsored Publications: Reports 124

5. U.S. Government-Sponsored Publications: Conference Papers 173

Name Index 201

Subject Index 207

PREFACE

This annotated bibliography of related federal and Commonwealth of Pennsylvania publications is designed to introduce researchers and students to significant government publications (hearings, reports and journal articles) and to government-sponsored research reports and conference papers related to the Three Mile Island nuclear generating station. Among the general topics covered are the development of the TMI Units 1 and 2; the 1979 TMI Unit 2 accident; the health-related issues and the economic, legal and social problems related to the TMI Unit 2 accident; the controversial issues and problems connected with the subsequent restart of TMI Unit 1, and the ongoing TMI Unit 2 clean-up and recovery activities.

     I am indebted to Mr. John Ganly, Chief, Economic and Public Affairs Division, Research Libraries of the New York Public Library, who first suggested the need for this bibliography. Ms. Jana Varlejs, Director of Professional Development Studies, School of Communication, Information and Library Studies, Rutgers University, rendered valuable assistance and encouragement for this project. Helpful advice and direction was given by Dr. Patricia Reeling, Chairperson, Department of Library and Information Studies, School of Communication, Information, and Library Studies, Rutgers University. Mr. John Geschwindt, Section Head, Government Publications, State Library of Pennsylvania, gave great help in locating essential government publications. Mr. Aaron Polonsky, Collection Development Librarian, Bloomsburg University, provided significant assistance in procuring essential background materials. Mr. Karl Proehl, Map Librarian, Pennsylvania State University, gave assistance in locating useful maps. My special thanks are extended to Mr. Doyle Dodson, Director, Office of Computer Services, Bloomsburg University and to Mr. Steven Boatman, Computer Operator, Bloomsburg University, for designing a program to facilitate the transfer of this bibliography onto computer disks.

Finally I wish to acknowledge the assistance of a host of librarians and friends. I would be remiss if I failed to acknowledge the value of the final typing of this manuscript extended to me through the Word Processing Center, Bloomsburg University. Mrs. Jane C. Harrison was responsible for the final typing of this manuscript. Any errors or omissions in this work are, of course, my sole responsibility.

                                      Peggy M. Hassler

Bloomsburg University
Bloomsburg, Pennsylvania
April 1988

INTRODUCTION

Until the April 26, 1986, accident at Chernobyl in the U.S.S.R., Three Mile Island (TMI) had the dubious distinction of being the scene of the world's worst commercial nuclear power plant accident. Russian announcement of the Chernobyl accident was delayed until that accident was several days old. By contrast, public notification of the TMI Unit 2 accident was made approximately four hours after serious nuclear reactor problems had begun. For some eight days from March 28 through April 4, 1979, the American news media presented an almost continuous account of the TMI Unit 2 accident and the efforts by plant operators and others to control the nuclear power emergency and to bring about a safe shutdown of the damaged nuclear reactor.

Radiation releases from the Chernobyl plant caused the evacuation of at least 135,000 persons who were living within an eighteen-mile radius of the crippled plant. Russian authorities have indicated the death of thirty-one persons who fought to contain the fires which began when the Chernobyl Unit 4 nuclear reactor exploded. A year after the Chernobyl disaster, large land areas in the U.S.S.R. remain contaminated. Russian news sources indicate that approximately 90,000 Chernobyl area residents have been permanently relocated in other areas of the country. Many foreign observers are skeptical that Russian authorities have disclosed full information concerning the health consequences of the Chernobyl accident.[1]

Fortunately, the TMI Unit 2 accident was contained within the building complex of the nuclear power unit. TMI Unit 1 was not affected by the TMI Unit 2 accident. Fortunately also, no lives were lost at Three Mile Island. Some eight years after the TMI Unit 2 accident, various health studies are still being conducted to determine possible adverse health effects upon local residents of the immediate area. Clean-up and recovery operations at the crippled TMI Unit 2 reactor are continuing, with the final TMI Unit 2 clean-up now projected for completion in 1988 or 1989.[2]

We may well ask what the causes of the TMI Unit 2 accident were. But before we discuss the accident and its causes, and relevant publications, we must understand the reasons which motivated U.S. utilities to develop very large commercial nuclear power plant complexes like the TMI Unit 1 and Unit 2 in the first place. We need to understand the advantages and problems of these U.S. utilities that use nuclear power to generate electricity. Finally we need to review briefly the commercial development of nuclear power in the United States, with particular attention to the incentives which prompted the development of the present-day commercial U.S. nuclear power facilities. Only then can the Three Mile Island accident be viewed in proper perspective.

Early U.S. Nuclear Power Research and Development

In the early 1930s serious nuclear fission research was begun in a number of countries with the purpose of using nuclear power to develop more powerful military weapons. Before World War II, German scientists and others had begun systematic research into the ways nuclear power could be used for armaments.[3]

Through the efforts of Enrico Fermi and his associates, the first sustained nuclear fission reaction was produced in the United States in 1942. Nuclear fission research was accelerated both in the United States and abroad during the war years. Ultimately, research by many scientists and technicians culminated in the U.S. production of two atomic bombs, successfully detonated over the Japanese cities of Hiroshima and Nagasaki, on August 7, 1945, and August 9, 1945, respectively.[4]

The two atomic bombs caused great devastation in both Japanese cities, with many thousands of fatalities and injuries. Undoubtedly the destruction caused by these bombs prompted the Japanese government to sue for peace a few days later. Many observers felt that the dropping of these bombs saved lives in the long run. The anticipated Allied invasion of the

Introduction 3

Japanese home islands was now rendered unnecessary and the war in the Pacific was shortened by a year or more.[5]

With World War II at an end, more peaceful uses of nuclear power could be considered. In 1945 the U.S. government controlled all nuclear fission research within the nation. Unless the federal government gave its express consent, no private or public corporation could develop any nuclear fission research program. The federal government retained the right to all products derived from any nuclear fission research. Even before World War II ended, scientists had become interested in the possible applications of nuclear fission for medicine and for the generation of electricity. Our attention will now turn to the federal legislation which assisted American utilities in developing commercial nuclear power facilities.

Federal Assistance for U.S. Commercial
Nuclear Power Development

Since World War II, the United States has had access to a continuing supply of uranium, the source of nuclear energy, and to nuclear fuels. With rising prices of oil, nuclear power advocates were quick to point out that nuclear power was a cheaper form of energy for American utilities than oil. In the 1980s nuclear power advocates champion the continued and expanded use of nuclear power to generate electricity as one way to reduce U.S. dependence upon foreign oil imports.

Following World War II, various federal laws were passed to assist U.S. utilities which wished to explore the possibility of using nuclear power to generate electricity. A 1983 government report, Nuclear Regulatory Legislation Through the 97th Congress, 2D Session (CIS 83 H442-1) summarized the texts of various nuclear power-related legislation enacted between 1954 and 1982.

Congress passed the McMahon Bill, better known as the Atomic Energy Act of 1946 (60 STAT 755) to permit the first non-military use of nuclear power in the United States. Provisions of this act created the Atomic Energy Commission (AEC), forerunner of the today's Nuclear Regulatory Comission (NRC). Like the NRC, the AEC was an independent federal commission, with members appointed by the President subject to Congressional approval. Other provisions of the act gave private industry the ability to investigate the uses of nuclear power for medically-related research and for the commercial generation of electricity. The federal government continued to retain complete control of all nuclear fission materials and products related to national defense and security.[6]

A Congressional Joint Committee on Atomic Energy was created by this 1946 legislation. This committee became the unofficial adviser for other Congressional committees on all matters pertaining to nuclear power. In time, this committee would influence the drafting of proposed nuclear power-related bills and the passage of nuclear-related legislation.

In the early 1970s a critic of the Atomic Energy Commission declared, "the JCAE is more knowledgeable about nuclear matters than are the AEC commissioners."[7] Since the JCAE was dissolved in 1977, its responsibilities have been shared by two committees. In the Senate, the Committee on Environment and Public Works created a special subcommittee on Nuclear Regulation to deal with nuclear power concerns. This committee would conduct an investigation of the TMI Unit 2 accident in 1979 for the U.S. Senate. In the House, the Committee on Interior and Insular Affairs authorized a special subcommittee to be responsible for nuclear power matters.

During the Spring of 1953, a small nuclear reactor at the federal Idaho National Engineering Laboratory produced electricity for the first time. Following this 1953 event, the Atomic Energy Commission authorized the development of a number of experimental nuclear power reactors at various federal testing centers. All of these reactors were small but they demonstrated that nuclear reactors could be used to produce electricity effectively. However, the federal government still retained ownership of all U.S. nuclear reactors and nuclear fuels. Additional incentives would be needed before American utilities would commit the research time and the capital necessary for the commercial development of large-scale nuclear power facilities for the generation of electricity.

American utilities received additional incentives for commercial nuclear power development with the passage of the Atomic Energy Act of 1954 (42 U.S.C. 2011 et seq.). All of the provisions of the Atomic Energy Act of 1946 (60 STAT.

Introduction 5

755) were incorporated into the new legislation. The broadened act gave the Atomic Energy Commission the power to license both federal authorities and private corporations to develop nuclear reactors for non-military purposes. Private corporations could now own nuclear reactors, provided these reactors were used only for non-military purposes. In addition, the AEC received federal funds for the development of nuclear reactors for non-military uses.[8]

In December 1953, the Atomic Energy Commission invited the Duquesne Light and Power Company of Pittsburgh, Pennsylvania and the Westinghouse Corporation to join the federal government in a unique three-way partnership to construct the first civilian U.S. nuclear power plant. Duquesne provided the land and $300 million, and guaranteed to operate and maintain the plant. Westinghouse, Inc. developed the reactor, based on a model then under development for the U.S. Navy's Submarine Propulsion Program. This plant at Shippenport, Pennsylvania, began commercial operation in 1957. Earlier, in 1954, the U.S.S. Nautilus, the first nuclear powered submarine, had been successfully launched.[9]

The future for U.S. commercial nuclear power plant development was bright with promise in 1954. The Power Demonstration Reactor Program (PDRP) was made possible by the 1954 Atomic Energy Act. This program was a joint undertaking by the federal government and private industry to develop nuclear reactors for commercial purposes. U.S. utilities could apply for federal assistance to develop nuclear power reactors for the commercial generation of electricity.

Project requests from U.S. utilities were submitted to the AEC describing proposed sites, equipment and engineering specifications, environmental considerations, costs, etc. All project requests were reviewed individually by the AEC. Each approved PDRP project had to be completed within a five-year period. Generally PDRP costs were apportioned as follows: private industry assumed seventy-five percent of the proposed costs, including personnel training and fuel fabrication; the federal government financed the remaining twenty-five percent of the project costs. Some fourteen nuclear power reactors were constructed through the PDRP program.[10]

Even though government incentives induced U.S. utilities

to begin research into the commercial uses of nuclear power to generate electricity, the problem of liability for damages in case of a serious nuclear power plant accident still remained. Without the passage of the Atomic Energy Damages Act of 1957, commonly known as the Price-Anderson Act (42 U.S.C. 2210), no large U.S. commercial nuclear power plants might have been constructed. The original legislation guaranteed public payment of up to $560 million for a single commercial nuclear power plant accident. This legislation mandated the establishment of a $60 million insurance fund, funded by those U.S. utilities using nuclear power to generate electricity.

Subsequent amendments to this legislation each time have raised this insurance fund to $160 million for a single nuclear power plant accident. Liability for a commercial nuclear power accident remains the responsibility of the parent corporation which owns the nuclear power facility at the time of a serious nuclear power accident. Each time Price-Anderson Act is due for reconsideration, its financial provisions have been increased. However, it must be noted that a guaranteed $560 million liability in 1957 seemed an adequate safeguard to take care of any possible commercial nuclear power plant emergencies.[11]

U.S. Nuclear Power Industry: Advantages and Problems

As previously indicated, U.S. utilities received federal assistance for research and development of commercial nuclear power facilities through various pieces of federal legislation. In addition to these laws, necessary capital was readily available to U.S. utilities at relatively low interest rates throughout the 1960s and into the early 1970s. The United States continued to have access to an ample supply of uranium and the cost of nuclear fuels remained lower than that for oil or coal. Throughout the 1950s, residential use of electricity grew at a rate of nearly ten percent annually. Industry forecasts indicated that the demand for electricity would continue to grow at a similar rate into the 1980s and 1990s. However, in the late 1960s the U.S. growth rate for electrical consumption dropped to seven percent annually and by the early 1970s U.S. electric sales figures showed less than a three percent increase annually. Industry forecasts which overestimated the U.S. consumption of electricity led to more nuclear power

Introduction

facilities being planned than might be needed. By 1970, United States electric utilities had eighteen operating nuclear power plants. In 1975, there were fifty-five nuclear power plants in operation in the United States and some one-hundred and seventy-eight were planned or were under construction.

In the 1960s American utilities became committed to the Light Water Reactor (LWR) design. Two general types of nuclear reactors were developed utilizing the general LWR design: the Boiling Water Reactor (BWR) and the Pressurized Water Reactor (PWR). The two Three Mile Island nuclear reactors used the LWR design and are PWR reactors. During this same time period, the U.S. Navy's Submarine Propulsion Program had adopted the LWR design for the nuclear reactors used in this endeavor. For many years Admiral Hyman G. Rickover directed this U.S. naval program. He became a strong advocate of the use of nuclear power for the commercial generation of electricity. Numerous former U.S. naval personnel, trained in the U.S. Navy's submarine propulsion program, found employment at the rapidly expanding U.S. commercial nuclear power facilities.[12]

The Three Mile Island Unit 2 accident shook public confidence in the general safety of nuclear power plant operations. Even before the TMI accident, cost overruns and lengthy construction periods had become problems for the American nuclear power industry in the 1970s. U.S. utilities did not adopt a common design model for the construction of nuclear reactors used in American power plants as the French nuclear power industry did. A French nuclear power facility can expect to have a five- to six-year construction period, whereas an American nuclear power plant may take twelve to fourteen years to become operational.

Keeping all of these factors in mind, together with the leveling off of U.S. commercial and residential needs for ever-growing amounts of electricity, it is not surprising that no new U.S. commercial nuclear power facilities have been authorized since 1978. Plants in operation or under construction at the time of the TMI Unit 2 accident have had to bring their facilities in line with new NRC safety recommendations for equipment and operator training made in the aftermath of TMI Unit 2 accident findings. Routine repair and maintenance costs have steadily risen at those U.S. nuclear power plants presently in operation.

A 1984 New York Times article suggests that the American commercial nuclear power industry expanded too rapidly, without allowing sufficient research time and funds to develop a truly safe nuclear reactor design. This article suggests that smaller, less complex nuclear power units might have taken precedence over the very large complex of U.S. nuclear power generating stations. The following example is cited to indicate the complexity of present-day nuclear technology: Approximately 40,000 valves are needed in a single large U.S. nuclear power plant, while an oil or coal-fired plant of a similar size requires only 4,000 valves.[13]

Since the TMI Unit 2 accident, many American nuclear power facilities have been cancelled, even though construction was well under way and considerable construction funds had been spent. Altogether some 114 nuclear power plants were cancelled in the United States between 1972 and 1984. Many of those units cancelled in 1983 and 1984 were abandoned when they were more than halfway to completion. Reasons cited for recent U.S. nuclear power plant cancellations included: cost overruns, construction delays, financial problems, safety issues, consumer opposition, and a reduced forecast for electrical sales for both commercial and residential use.[14]

Now we are ready to consider the development of the Three Mile Island nuclear power facility, the TMI Unit 2 accident, and the subsequent clean-up and recovery operations, together with related government publications and selected government-sponsored research reports and conference papers.

Three Mile Island: Nuclear Power Plant Development

Three Mile Island is located in the Susquehanna River in Dauphin County, Pennsylvania. The island lies approximately ten miles southeast of Harrisburg and about three miles south of the borough of Middletown. In 1979 Middletown had a population of 10,000 persons, while the greater Harrisburg area had a population of 230,000 individuals. At the time of the TMI Unit 2 accident, a Washington Post article indicated that 24,527 persons were living within a five-mile radius of Three Mile Island, another 133,672 persons were living within a ten-mile radius and a grand total of 636,073 persons had homes within a twenty-mile radius of Three Mile Island.[15]
Along the sixty-mile stretch of the Susquehanna River between

Introduction

Harrisburg and Conowingo, Maryland, there are nine electrical generating stations of diverse types--hydroelectric, pumped storage, coal and steam, and nuclear. Together these various stations have a combined generating capacity of over 5,000,000 kilowatts to serve the potential power needs of more than 2,000,000 customers.[16]

Three Mile Island is small. It contains some 625 acres and is approximately three miles long--hence the name. As the most recent Ice Age glaciers receded, this island was formed. Large boulders and debris were deposited in the shallow channel near the Susquehanna River's eastern bank. Very gradually river silt collected around the boulders and slowly the island evolved to its present size.[17]

Before construction was begun for TMI Unit 1, the Pennsylvania Historical and Museum Commission received permission to conduct a series of archeological excavations on Three Mile Island to determine what traces of settlement had been left by previous cultures. These archeological investigations took place during the summer of 1967. Findings indicated that the island had been used for hunting and fishing for more than 9,000 years. Fragments of soap bowls and clay pottery were found which can be dated as early as 1700 B.C. The most recent Native American inhabitants of this region were the Susquehannock Indians. Large numbers of arrowheads, spear points and miscellaneous stone tools were discovered and identified as having been produced and used by the Susquehannocks. However, the archeological excavations failed to uncover evidence of any permanent settlements on the island by previous cultures.[18]

In 1906, Three Mile Island was purchased by the York Haven Water and Power Company, now a subsidiary of Metropolitan Edison Company. For the next sixty years, the island changed very slightly. Farmers leased about half of the land to grow corn and tomatoes. The cornfields were frequently visited by ring-necked pheasants which found both food and a resting place here. A picnic area and a boat landing site were developed and maintained by the power company for public use. Near the southern end of the island, a stand of black locust trees offered a shelter to deer and other wildlife. Today only the trees remain, dwarfed by four huge 370-foot cooling towers for the Three Mile Island nuclear power plant.[19]

By November 1966, the General Public Utilities Corporation, the parent company for Metropolitan Edison, and its associated companies had selected Three Mile Island as the potential site for construction of a nuclear power unit. Public announcement of this decision was made in the New York Times and in local Pennsylvania papers on November 19, 1966.

Economic considerations played a major role in this decision. As Three Mile Island was already owned by Metropolitan Edison, no land purchase was necessary. The parent corporation, General Public Utilities, Inc., was satisfied that the necessary cooling water for the proposed TMI facility could be obtained at a lower cost from the Susquehanna River near Three Mile Island than at any of the other possible nuclear facility sites then under consideration. Power transmission lines were already in place and could be utilized as the older, less-profitable coal-fired plants were retired from service. Adequate transportation facilities existed nearby to bring in necessary equipment and supplies to construct the nuclear power facility. Similarly, these same railroad lines and highway road systems could provide the means to ship the anticipated waste water from the plant to approved nuclear waste depositories once the nuclear power plant was in operation.[20]

In 1966 there was no widespread local opposition to the construction of a nuclear power plant so close to a large population center. Few area residents were concerned about the potential dangers of a nuclear accident. Rather, numerous residents viewed the proposed construction at TMI as a means of improving local employment opportunities and increasing tax revenues. With the anticipated closing of Olmstead Air Force Base in 1967, some local jobs would undoubtedly be lost and many residents believed the proposed TMI construction project could take up the economic slack.[21]

Following the necessary approval by the Atomic Energy Commission, construction of TMI Unit 1 was begun in 1968. Gilbert Associates, Inc. served as architect-engineer for the project. Babcock and Wilcox supplied the massive nuclear reactor and its component parts, and General Electric supplied the turbine generator. Construction costs for TMI Unit 1 were approximately $400 million. On September 30, 1974, TMI Unit 1 began commercial operation.[22]

During 1968 General Public Utilities Corporation and its

Introduction   11

subsidiaries agreed on the feasibility of constructing a second nuclear power unit at Three Mile Island. Prior to this decision, another possible site had been considered on Oyster Creek, New Jersey. Metropolitan Edison suggested in its application to the Atomic Energy Commission, that the TMI site was "economically advantageous." The various advantages noted in the original TMI Unit 1 application were reconfirmed. This application indicated that the Harrisburg area had a good supply of construction workers who could be hired at lower hourly rates than was possible at other potential construction site areas. Two area businesses, the Hershey Chocolate Corporation and the Steelton unit of Bethlehem Steel, were mentioned as important customers for large amounts of electricity.[23]

Following approval by the Atomic Energy Commission, construction of TMI Unit 2 began in 1970. Burns and Roe, Inc. served as the architect-engineer for this project. Again Babcock and Wilcox supplied the nuclear reactor and its component parts, while Westinghouse, Inc. furnished the turbine generator. In contrast with Unit 1, TMI Unit 2 took nearly nine years to construct and cost $700 million, nearly double the cost of TMI Unit 1. On December 30, 1978, TMI Unit 2 began commercial operation.[24]

Until the 1979 TMI Unit 2 emergency, Metropolitan Edison operated both TMI units for the benefit of General Public Utilities Corporation and its associated companies. Following the TMI Unit 2 accident, GPU, Inc. created another unit, GPU Nuclear, to take direct responsibility for the Three Mile Island Nuclear Generating Station. The three subsidiary companies of GPU, Inc. continue to own the TMI Nuclear Generating Station in these proportions: Jersey Central Power and Light (JCP&L), twenty-five percent; Metropolitan Edison Company (Met-Ed), fifty percent; and Pennsylvania Electric (Penn-Elec), twenty-five percent.[25]

Both of the Three Mile Island units have nuclear reactors of the Light Water Reactor (LWR) design. This type of reactor is so named because ordinary water is used as a coolant or medium to transfer heat. As mentioned before, there are two types of Light Water Reactors (LWR's): the Boiling Water Reactor (BWR) and the Pressurized Water Reactor (PWR). In the BWR, heat from the nuclear fussion causes the water to boil in the reactor vessel and the water is turned into

FISSION

steam. Then this steam is pumped directly to the turbine. [handwritten annotation: "STEAM IS NOT PUMPED"] In the PWR, water is pressurized to prevent its boiling in the reactor vessel. Then this pressurized, or primary system water, is pumped through a steam generator, or heat exchanger, where the heat transfer takes place and the water in the secondary system is brought to a boil. The steam thus produced drives the turbine.[26]

At the Three Mile Island Generating Station, each of the two pressurized water nuclear reactors is located in its separate huge containment building. Each TMI containment building was designed so as to prevent the spread of radiation in case of a serious nuclear power accident. The TMI Unit 2 containment building housed the extremely large cylindrical nuclear reactor core, a steel vessel containing some 177 hanging fuel assemblies. In each fuel assembly of TMI Unit 2, there were some 208 enriched uranium pellets, making a total of 36,816 uranium pellets within the reactor.

Control rods separated the fuel assemblies; these rods absorbed the neutrons and acted as a vehicle to vary the fission rate. As the control rods were raised out of the fuel assemblies, more neutrons became available to split the uranium atoms and to produce additional heat. When the control rods were lowered into the fuel assemblies, they absorbed neutrons, reduced fission, and lowered the temperature in the reactor core. When an operational problem within the nuclear reactor developed, the control rods were programmed to lower into the fuel assemblies and to shut down the reactor automatically. For a more detailed description of Three Mile Island's Unit 1 and Unit 2, the reader may wish to consult chapter two of <u>Crisis Contained: The Department of Energy at Three Mile Island</u>.[27]

Since the pressurized water reactor is such a complex mechanism, there are numerous opportunities for operational problems. Presently the Nuclear Regulatory Commission, the successor of the Atomic Energy Commission, has the responsibility for licensing and for the operational safety of all U.S. commercial nuclear reactors. All operational problems within a commercial nuclear facility are classified as "events" by the NRC. During the licensing and early operating stages of an American commercial nuclear power plant owner, the appropriate nuclear plant owner must report promptly all "events" to the Nuclear Regulatory Commission.

Introduction 13

The final Safety Analysis Report for TMI Unit 2 lists various possible "events" which might take place, ranging from the malfunctioning of valves to the hazards of a serious earthquake. Nowhere in this report is there mention of the potentially serious problem which a clogged demineralizer might cause. The demineralizer is mentioned incidentally but its continued operation is not considered essential to the safe shutdown of the nuclear power unit.[28]

At TMI Unit 2, on March 28, 1978, an electromatic relief block valve failed to close for four minutes after a fuse had blown. On this occasion the reactor tripped and the emergency coolant system functioned as expected. Later, plant operators recalled that they did not understand the cause for the coolant systems's sudden depressurization. The TMI Unit 2 control room had no indicator to show whether the block valve was open or closed. During subsequent testimony about the TMI Unit 2 accident, plant operators voiced their concern that "the block valve could stick open or shut if it was used too often."[29]

Prior to licensing, the Three Mile Island Unit 2 experienced a series of difficulties involving pump and valve failures. All of these problems were reported to the NRC but none was considered serious enough to delay the commercial licensing of TMI Unit 2 on December 30, 1978.

Later TMI Unit 2 accident investigations would censure the NRC, TMI Unit 2's managers, and the Babcock and Wilcox Company for failing to give sufficient attention to the possible consequences of pressurizer valve failures. Indeed, pressurizer valve problems were not unique to Three Mile Island Unit 2. A potentially serious accident in 1977, involving failure of the same type of pressurizer valve which stuck open at TMI Unit 2, had occurred at the Davis-Besse nuclear power plant in Toledo, Ohio. Over a thousand gallons of water spilled into the Davis-Besse's containment building before the open block valve was bypassed and the flow of water halted. Several Babcock and Wilcox engineers were sufficiently alarmed to write detailed memoranda about this accident, nearly a year before the TMI Unit 2 emergency. These engineers urged immediate improvements in the design of the pressurizer valve and correctly indicated that premature shutoff of emergency valves and pumps might mislead plant operators during a nuclear power plant emergency.[30]

In the first quarter of 1979, various operational problems continued to occur at TMI Unit 2. On February 7, 1979, a diaphragm on the waste tank rupture disk broke and it had not been replaced before the TMI Unit 2 emergency began. In the control room, the alarm printer was supposed to deliver a record of reactor problems for plant operators. However, TMI Unit 2's alarm printer had a record of frequent jamming, causing paper to tear and to produce illegible records. Frequently this alarm printer ran too slowly to assist plant operators. During the TMI Unit 2 accident's early stages, the alarm printer malfunctioned and provided no assistance for the TMI Unit 2 plant operators.

On March 26, 1979, a maintenance crew conducted a routine test of the valves in TMI Unit 2's emergency cooling system. As part of the test, the valves were closed temporarily to the emergency cooling system. During subsequent accident investigations, these men testified that these valves had been reopened following this test. However, on March 28, 1979, these same pressurizer valves remained closed and water did not reach the Unit 2 reactor core. The maintenance crew would later testify, " ... the pressurizer valves did not always reopen when they were supposed to do so."[31] An NRC official, Harold Denton, would later suggest, " ... pressurizer valves had stuck open approximately one hundred times on reactors developed by Babcock and Wilcox...."[32] This statement was based on a tabulation of "events" reported to the NRC by appropriate U.S. commercial nuclear power facilities.

Now we can turn to the TMI Unit 2 accident development.

Three Mile Island: Accident at Unit 2

On Wednesday, March 28, 1979, Three Mile Island Unit 1 reactor was shut down for the loading of nuclear fuels and for a routine unit inspection. In the early morning hours, only TMI Unit 2 was in operation. The plant operators on duty during this shift were all former U.S. Navy personnel who had received their initial introduction to the operation of nuclear reactors through the U.S. Navy's Submarine Propulsion Program, then directed by Admiral Hyman G. Rickover.[33]

Introduction   15

For the past eleven hours, maintenance crews had tried
unsuccessfully to clean out TMI Unit 2's clogged demineralizer
system. This system was important to power plant operations
as it separated river salts and silt from the river water before this liquid entered the reactor's pipes. If the water
could not flow freely through the reactor's pipes, the entire
system was programmed to shut down. Successive crews had
used high pressure hoses containing a mixture of air and water
in an attempt to clean out the demineralizer system. Finally,
about 4:00 a.m., the TMI Unit 2 demineralizer became completely clogged, the mainwater pumps tripped and the turbine
shut down. In response to the above-mentioned actions, the
auxiliary feedwater pumps started automatically as programmed.
The TMI Unit 2 accident was about to begin.

Even though the auxiliary feedwater pumps started automatically, no water was reaching the steam generator. As
previously mentioned, valves in the line to the auxiliary feedwater pumps had been closed during a recent routine system
test. Unintentionally these valves remained closed and now no
feedwater was reaching the steam generator. Plant operators
were misled by their instruments, which indicated that water
was rising in the reactor's core. It was some eight minutes
before plant operators realized that these valves were shut
and opened them manually.³⁴

Although the reactor pressure dropped as anticipated,
an electromagnetic pressurizer relief valve (PORV) remained
stuck open, eventually permitting reactor coolant water to
spill into the Unit 2 containment building. It is still a matter
of conjecture as to why this valve stuck open. The prior
warning of possible failure of this same type of pressurizer
valve, given after the earlier Davis-Besse plant emergency,
had not been heeded.

Both mechanical and human errors continued to increase
TMI Unit 2's problems. Two minutes after the TMI Unit 2
accident began, the emergency core cooling system started up
as programmed. A minute later, a plant operator, misled by
his instrument gauges, turned off this cooling system, and
so turned back the needed flow of water to the nuclear reactor. Plant operators were concerned, mistakenly, about there
being too much water in the system and the resulting loss of
pressure. In reality the plant operators should have been

concerned about the dangerous loss of water and possible uncovering of the reactor core.[35]

Problems multiplied rapidly. A false fire alarm sounded in TMI 2's control room, adding to the confusion. Paper repeatedly jammed in the alarm printer, rendering it almost useless to the plant operators. Within the Unit 2 containment building, radiation levels increased alarmingly. Plant operators became increasingly concerned about reactor fuel damage and leakage.

About 6:00 a.m., when the TMI Unit 2 accident was about two hours old, plant operators arranged a conference call to the homes of several company officials to discuss the mushrooming crisis. Later, a summary of this thirty-eight-minute call indicated a consensus to trust the TMI Unit 2 instrument readings. In the course of the call, plant operators indicated that there had been no radiation releases outside the Unit 2 containment building. Unknown to the plant operators, TMI Unit 2's reactor core must have become uncovered sometime during this conference call.[36]

Between 6:24 a.m. and 6:35 a.m., TMI Unit 2 radiation monitors began to sound alarms, indicating excessive radiation levels. Meanwhile, plant personnel began surveying radiation levels within the plant. About three hours after the TMI Unit 2 accident had begun, a Site Emergency was announced at 7:02 a.m. The Pennsylvania Emergency Management Agency was informed by phone of the TMI Unit 2's escalating problems and the operator on duty was requested to contact the Pennsylvania Bureau of Radiation Protection with this information.[37]

Continued high radiation readings prompted the announcement of a General Emergency about 7:30 a.m. The Nuclear Regulatory Commission's Regional Office near Philadelphia and the Civil Defense Office in Harrisburg were notified of the growing crisis. Later, state and federal officials would sharply criticize Metropolitan Edison for its delay in notifying appropriate government agencies and officials about the rapidly developing emergency at TMI Unit 2.

Around 8:00 a.m. plant operators realized that the core temperatures had reached 2620°F. Plant operators recognized the possibility that the reactor core could have been uncovered.

Introduction 17

There was grave concern about possible damage to TMI Unit 2's zircalloy fuel cladding, the probable release of fission products, and the possible generation of hydrogen.[38]

Rapidly the news of the TMI Unit 2 accident reached the outside world. Governor Dick Thornburgh gave a brief radio announcement concerning the TMI Unit 2 emergency at 10:25 a.m., advising people living within ten miles of the plant to stay indoors and to keep doors and windows closed. The governor considered ordering an immediate evacuation of the Harrisburg area because of the rapidly escalating crisis, but was persuaded to defer this plan for the present. For some hours, both Metropolitan Edison officials and plant operators thought the risk of radiation releases were diminishing.[39]

Assistance came to the crippled nuclear power facility from a variety of government agencies and departments and from nuclear experts in all parts of America. The Nuclear Regulatory Commission and the Pennsylvania Department of Environmental Resources promptly sent teams to the site to begin radiation readings. Department of Energy personnel set up an Emergency Operations Center in an empty Pennsylvania Department of Transportation hangar at the Capital City Airport at New Cumberland, Pennsylvania.[40] Within the troubled plant, the ventilation of TMI 2's auxiliary building was begun. Radiation levels continued to be high within TMI Unit 2's containment building throughout the first day of the TMI accident. Plant operators indicated that the presence of the "hydrogen bubble" in the TMI Unit 2 reactor represented a threat to a safe shutdown.[41]

As TMI Unit 2's problems increased, the NRC sent Harold Denton, then Director of the Office of Nuclear Reactor Regulation, to the TMI Unit 2 site. He served as spokesman for the NRC at the frequent news conferences held during the next few days. Babcock and Wilcox engineers, various industry and utility teams, teams from a variety of state and federal agencies, and scientists and engineers from the United States and abroad rapidly converged on the TMI Unit 2 site to assist in bringing the crippled plant to a safe shutdown.[42]

Some 400 news media representatives rapidly joined the contingency of nuclear power experts at Three Mile Island. It was agreed that news coverage of the TMI Unit 2 emergency, on the whole, was careful and responsible. However, conflicting news reports added to the general uncertainty about the severity of the TMI accident and the progress being made towards safely shutting down the damaged TMI Unit 2 reactor. Reports from Metropolitan Edison and the NRC were frequently in disagreement. Finally, on April 1, 1979, President Jimmy Carter resolved this communication problem by ruling that henceforth all further technical reports on the TMI Unit 2 emergency should come from the NRC.[43]

Again Governor Thornburgh was faced with the decision of whether or not to order a general area evacuation. Throughout the TMI Unit 2 crisis, the responsibility for ordering a a general area evacuation would remain a paramount concern for the governor. On Friday, March 30, 1979, the governor ordered schools in the vicinity of TMI closed. All government offices in Harrisburg closed early that day. Governor Thornburgh requested all persons living within the vicinity of Three Mile Island to remain indoors and urged all pregnant women and preschool children, living within five miles of the TMI facility, to leave the area for the duration of the crisis.[44]

Rapidly Governor Thornburgh's emergency management team coordinated plans for a possible evacuation of all residents living within a twenty-mile radius of Three Mile Island. During this Friday, an evacuation was seriously considered as the hydrogen "bubble" impeded efforts at the crippled power facility to cool down the damaged reactor. Many residents did not wait for a possible evacuation directive; they packed their cars and left the Harrisburg area with their families without panic or fanfare. Estimates varied greatly, but between March 28 and April 3, 1979, it is believed that more than 144,000 persons left the area temporarily.[45] There were no massive traffic tie-ups or mass hysteria. Later surveys indicated that the vast majority of temporary evacuees went to stay with relatives or friends. The American Red Cross set up a temporary evacuation center in Hershey but received only small numbers of the area evacuees.[46]

President Carter visited the troubled TMI Unit 2 site on Sunday, April 1 and toured the TMI Unit 2 facility with Governor Thornburgh and various nuclear power experts. The President's visit was interpreted by the news media as a means of reassuring area residents that the TMI Unit 2 accident was being dealt with competently but also was preparing local residents for a calm, well ordered evacuation, should circumstances dictate one. As events turned out, no general area evacuation was necessary.

Later on in the evening of April 1, 1979, Governor Thornburgh announced that all government offices would be open for business as usual on Monday, April 2.[47] An undisclosed number of area businesses remained closed that Monday. In the course of his Monday news conference, Governor Thornburgh said he was pleased to announce that the hydrogen "bubble" was decreasing in size and a general area evacuation was no longer an immediate possibility. Finally, on Wednesday, April 4, 1979, Harold Denton assured another news conference that the danger of a hydrogen explosion within TMI Unit 2 reactor was now over. On April 9, Governor Thornburgh's advisory for pregnant women and preschool children was lifted.[48]

Preliminary environmental data collected after the inception of the TMI Unit 2 accident were also encouraging. None of the 130 water samples taken by the Nuclear Regulatory Commission, the Department of Energy or the Commonwealth of Pennsylvania between March 29, 1979, and April 3, 1979, showed any traces of radioiodine. Again, no radioiodine was found in some 147 soil samples taken by the NRC or the DOE. Similarly 171 vegetation samples showed no traces of radioiodine. Milk samples tested by the state showed no traces of radioiodine, but nine milk samples tested by the Federal Food and Drug Administration did show some slight traces of radioiodine.[49]

In the course of draining the contaminated water from TMI Unit 2's auxiliary building, three plant workers were exposed to radiation levels which exceeded the three rem per quarter standards recommended by nuclear safety guidelines. At least twelve other TMI workers were exposed to radiation levels below the three rem per quarter as they assisted in removing the contaminated water. Continued releases from TMI Unit 2's industrial system triggered considerable opposition long after TMI Unit 2 crisis had ended.[50]

By most definitions, Three Mile Island cannot be classified as a disaster like the 1986 U.S.S.R. accident at Chernobyl. At TMI Unit 2 no lives were lost and serious environmental problems were confined to the nuclear generating station site. At a news conference on April 3, 1979, Governor Thornburgh declared: "Nuclear opponents who would shutdown every nuclear reactor in the country tonight simply are not in touch with reality. But nuclear advocates who would pretend that nothing is changed by our vigil at Three Mile Island, simply are not in touch with reality."[51]

Three Mile Island: Accident Investigations and Related Government Publications

Even before the crippled Three Mile Island Unit 2 reactor had been brought to a safe shutdown, there was widespread public demand for a full inquiry into the causes and effects of the accident. Pennsylvania Senator Richard Schweiker was among the first to call for the appointment of a special Presidential commission to investigate the TMI accident and recommend what future role nuclear power facilities should play in the generating of electricity for the United States.

Meanwhile General Public Utilities, Inc. suspended all construction work on its various new facilities and curtailed all but the most essential construction activities at its operating plants. The liability insurance guaranteed by the Price-Anderson Act soon proved inadequate to provide TMI Unit 2 clean-up and recovery costs. Massive clean-up costs at TMI Unit 2 became an immediate financial burden for GPU and its associated companies. Replacement power costs to compensate for the loss of electricity generated at the Three Mile Island site would be an additional concern for GPU until TMI Unit 1 could be restarted.[52]

Following the TMI Unit 2 emergency, the Nuclear Regulatory Commission directed those nine U.S. nuclear power generating stations, operating with Babcock and Wilcox reactors similar to the TMI Unit 2 model, to immediately review their reactor emergency procedures and conduct complete inspection of their nuclear reactors within a ten-day time frame. Eight of the nuclear facilities with Babcock and Wilcox reactors, similar to TMI Unit 2, were permitted to resume operations following a subsequent NRC review. Only Three Mile Island Unit 1 was not cleared for continued operation, and so the restart of TMI Unit 1 became an important goal for the General Public Utilities Corporation.[53]

To facilitate its own investigation of the TMI Unit 2 emergency, the NRC set up a special review group. Subsequently the U.S. Senate conducted its own inquiry into the

Introduction   21

TMI Unit 2 accident, utilizing the services of the Subcommittee on Nuclear Regulation, a subcommittee of the Senate's Committee on Environment and Public Works.

A fourth important report on the consequences of the Three Mile Island accident was made by a specially appointed Commonwealth of Pennsylvania Commission, appointed by Governor Thornburgh and chaired by Lt. Governor William W. Scranton III.

All of the four major investigations held hearings, digested the testimony of numerous witnesses, and ultimately issued lengthy published reports. Since the Pennsylvania inquiry was primarily concerned with the consequences of the TMI Unit 2, it will be described following the various federal investigations.

President Carter appointed a special twelve-member commission to investigate the Three Mile Island accident and to make general recommendations concerning the future role of nuclear power in generating electricity in the United States.[54] Dr. John Kemeny served as the commission chairman. A distinguished nuclear expert, he had served as president of Dartmouth College since 1970. Congress acted quickly to approve the various commission members and to grant this presidential commission special powers to hold hearings and to subpoena witnesses. Frequently the commission is referred to as the Kemeny Commission, but its official title was the President's Commission on the Accident at Three Mile Island. The complete title of its published report is <u>The Report of the President's Commission on Three Mile Island: The Need for Change: The Legacy of TMI</u> (MC80-6415). Staff reports of subsequent investigations authorized by the Kemeny Commission were also published.

The Kemeny Commission investigated both the causes and the development of the TMI Unit 2 emergency. This report contained detailed descriptions of the various problems which occurred as the TMI accident progressed, together with recommendations as to how these crises might have been handled more effectively. Recommendations were made to facilitate improvements in emergency management training for all U.S. nuclear plant operators. The installation of improved nuclear reactor equipment, specifically designed to prevent problems which occurred at TMI Unit 2, was recommended.[55] Another recommendation suggested the restructuring of the NRC, with the proposed creation of a single executive head to replace the present commission organization. This last proposal was made because a majority of commission members believed a single executive might exert a greater degree of control over U.S. commercial nuclear power facilities than NRC had done up to that time.[56]

The Kemeny Commission did not recommend a moratorium of all U.S. nuclear power plant construction, as many critics of nuclear power had expected. But the Commission did recommend that any new U.S. nuclear power plant construction should take place at some distance from population centers.

Human error, flaws in the basic nuclear reactor design and a number of mechanical problems with the TMI Unit 2 facility were indicated as the immediate causes of the TMI Unit 2 accident. Both General Public Utilities Corporation and Metropolitan Edison Company were cited for failure to provide adequate emergency training for plant operators. Babcock and Wilcox was cited for not promptly correcting certain design flaws in its nuclear reactor design and in overall control room planning.[57]

The report suggested the NRC needed a much stronger policy for the regulation of U.S. utilities using nuclear power to generate electricity if future nuclear power plant accidents were to be avoided.[58] In conclusion, the Kemeny Commission indicated that a combination of causes was responsible for the TMI Unit 2 accident: inadequate training in emergency procedures for nuclear power plant operators; special problems in control room design, and design flaws within the nuclear reactor; and the combined failure of the Nuclear Regulatory Commission, the General Public Utilities, Inc. and its subsidiaries, and the Babcock and Wilcox Company to learn necessary safety lessons from previous U.S. nuclear power plant accidents, in particular the 1977 Davis-Besse nuclear power plant accident.[59]

As previously mentioned, the Nuclear Regulatory Commission utilized a Special Inquiry Group to develop a detailed analysis of the TMI Unit 2 accident and its causes and to recommend appropriate changes in NRC supervision of American nuclear power facilities. To avoid any charges of im-

Introduction 23

propriety, the NRC hired an independent Washington, D.C. law firm to conduct the NRC's investigation. Mitchell Rogovin, head of the law firm and George T. Frampton, Jr., a law firm partner, coordinated the Special Inquiry Group's activities.

The published NRC report has often been referred to as the Rogovin Report. The complete title of this several-volume report is Three Mile Island: A Report to the Commissioners and to the Public (MC 80-15452). These volumes are invaluable for any in-depth research of the TMI Unit 2 accident. Because of the wealth of information in these volumes, only a few important examples can be cited. Volume I, part one, included detailed information about TMI Unit 2 prior to the 1979 accident. Volume II, part two, contained a lengthy description of TMI Unit 2 accident development, and volume II, part three focused on the response to the TMI Unit 2 emergency by various groups, including the TMI Unit 2 plant operators; the GPU Corporation and its subsidiaries; and various local, state and federal agencies and departments. Numerous recommendations are presented which could be used to improve power plant safety and emergency planning at all operating U.S. nuclear power facilities as well as those under construction.[60]

A third investigation of the TMI Unit 2 emergency was undertaken under the auspices of the Senate Subcommittee on Nuclear Regulation. As previously mentioned, this subcommittee is one of two Congressional committees which took on the duties of the former Congressional Joint Committee on Atomic Energy (CJAE) in the 1970s. Senator Gary Hart (Democrat, Colorado) served as the subcommittee chairman during the time the subcommittee was engaged in its TMI Unit 2 accident investigation. For this reason, the published report has been referred to often as the Hart Committee Report. This Senate report is titled: Nuclear Accident and Recovery at Three Mile Island, A Special Investigation (MC 81-6337 and CIS 80 S 322-12).

The Hart Committee focused its attention on the consequences of the TMI Unit 2 accident for the nation as a whole. This report contains much information for the beginning researcher. For example, there is a detailed explanation as to how a nuclear power plant operates.[61] A useful summary of the various financial, health, legal, and social problems which have developed as a consequence of the TMI Unit 2 accident is also included.[62]

In summary, all of the three federal investigations of the 1979 TMI emergency made similar judgments about the TMI Unit 2 accident and its causes. The various reports agreed that plant operational errors compounded the TMI accident's severity but concluded that the plant operators were misled by flaws in nuclear reactor design, faults in instruments and inadequate emergency management training. All of the investigations concluded that the experiences at TMI Unit 2 pointed to the need for more adequate evacuation planning procedures for all population centers near commercial nuclear power facilities. Finally, all three investigations stressed the need for greater cooperation and shared research among the NRC, the U.S. nuclear power industry, and appropriate local, state, and federal departments and agencies to prevent future American nuclear power plant accidents.

As previously noted, the Commonwealth of Pennsylvania conducted its own investigation of the effects of the Three Mile Island accident upon Pennsylvania's residents, with particular attention to problems of those persons living within the vicinity of TMI Unit 2. Pennsylvania's Governor Thornburgh appointed a special fourteen-member commission composed of government officials and private citizens to complete a threefold task: to assess the actions of Commonwealth departments and agencies during the TMI Unit 2 crisis; to evaluate the TMI emergency's consequences for local Pennsylvania residents; and to make necessary recommendations for the improvement of Pennsylvania's emergency preparedness programs.[63] Lieutenant Governor William Scranton III, son of Pennsylvania's well-known former Governor William Scranton, was appointed chairman of this commission. Prior to this appointment, Lt. Governor Scranton had had responsibility for overseeing Commonwealth agencies concerned with energy programs and energy management. When this Pennsylvania commission completed its investigations it issued a report titled <u>Report of the Governor's Commission on Three Mile Island</u> (PYT 5312.2: R311G).

This study provided necessary background materials so that a beginning researcher can evaluate the numerous government publications dealing with the TMI Unit 2 emergency. A careful examination of this report is recommended for any researcher attempting to evaluate the consequences of the TMI Unit 2 accident. This report reviewed the economic, environmental, and legal problems which developed as a consequence

Introduction 25

of the emergency. Detailed summaries of proposed health-related studies are described which pertain to Pennsylvania residents living within the vicinity of the damaged TMI Unit 2 reactor. One useful appendix presented the Commonwealth's 1978 Disaster Operations Plan, while another described the Commission's recommendations for improved emergency preparedness planning made in the light of the TMI Unit 2 emergency.[64]

Three Mile Island: Clean-up, Recovery and Related Government Publications

Eight years after the TMI Unit 2 accident began, clean-up and recovery operations are still under way at TMI Unit 2. Following the TMI emergency, the General Public Utilities Corporation created a new administrative unit, GPU Nuclear, to deal with the enormous problems of the TMI accident. Immediately following the TMI Unit 2 emergency, activities were concentrated on assessing the accident damage and determining the procedures and systems best designed to deal with the crippled reactor.

Even though the condition of the TMI Unit 2 reactor could not be determined precisely, it was clear that massive fuel damage had occurred during the accident. The TMI Unit 2 reactor building was inaccessible due to high radiation and radioactive contamination levels and most of the complex of auxiliary and fuel handling buildings was almost entirely inaccessible as well. Many hundreds of thousands of gallons of highly contaminated radioactive water had spilled into the TMI Unit 2 reactor building basement and into the auxiliary building tanks. The first necessity was to stabilize conditions so that safe recovery operations could proceed. TMI Unit 2's immense technical difficulties were compounded by public anxiety, regulatory uncertainty, funding problems and the subsequent closing of commercial nuclear waste disposal sites to TMI Unit 2 wastes.[65]

Early recovery work included a number of programs and procedures designed to deal with technical clean-up activities. New types of plant systems were needed, including several decay heat removal systems, a nuclear sampling system, a pressure and inventory control system and various instrument and control systems. Liquid waste processing systems had to be designed and facilities constructed to store safely solid and liquid radioactive wastes for an extended time period. Finally the auxiliary and fuel handling building had to be decontaminated while planning was begun in preparation for reentry of TMI Unit 2's containment building.[66]

Recovery work at Three Mile Island Unit 2 was focused initially on stabilizing plant conditions so that the TMI Unit 2 auxiliary complex and the containment buildings could be safely entered. To gain access to these areas, the venting of krypton gases was proposed, which provoked considerable public opposition and hostility from anti-nuclear activists. Governor Thornburgh requested the Union of Concerned Scientists, a well-known organization of nuclear power critics, to study the design for the proposed venting plan and to make recommendations to him concerning public health and safety. After this group's research findings concluded that the proposed venting would not endanger the public, the krypton venting was allowed to proceed.[67]

Massive clean-up at the TMI Unit 2 complex became a great challenge for the entire U.S. nuclear industry. At TMI Unit 2 engineers searched in vain for a floatable water pump before purchasing an ordinary well pump and adapting it with a makeshift styrofoam life preserver. This contraption, called by its creators "S.S. Sumpsucker," successfully pumped out most of the 600,000 gallons of radioactive water from TMI Unit 2's containment building.[68]

In an interview with Wall Street Journal analysts, Paul Bengel, nuclear engineer overseeing the TMI Unit 2 clean-up, and Doug Bedell, GPU spokesman, indicated that adaptation and innovative design have been used throughout the eight-year TMI Unit 2 clean-up.[69] Radioactive particles have been blasted from building walls with powerful water hoses. Next, lawn mower-like "scabblers," adapted from an industrial cleaning device, and jackhammers have been used to remove several inches of the badly contaminated floors. Workers have used large wands to steam-clean radiation from overhead building pipes. The Westinghouse Electric Corporation devised a mixture of Tide detergent and Windex cleaner, common household products, which has successfully removed radioactive coating from the walls of TMI Unit 2's buildings.[70]

Plant workers, engaged in decontamination projects like TMI Unit 2, must wear protective clothing to reduce the dan-

Introduction 27

gers of radioactive exposure. These garments are bulky and resemble a cross between a space suit and an Eskimo's clothing. In the "hottest" areas, those with the highest radiation levels, four separate levels of clothing are required. Here all decontamination workers must wear vinyl suits over two layers of cloth overalls and suits of paper-like materials. Usually workers spend no more than two to four hours at a time working in heavily contaminated plant areas, depending upon radiation levels.[71]

Throughout the TMI Unit 2 clean-up and recovery, new equipment and procedures have had to be devised to solve various problems. A high polar crane had to be developed to recover radioactive debris from the TMI Unit 2 reactor's core. Once this polar crane was in operation, another problem complicated the removal of radioactive wastes. Algae, formed in the radioactive water surrounding the partially covered reactor core, obscured workers' vision. Since the nuclear engineers feared that ordinary pool cleaners, like chlorine, might corrode TMI Unit 2's reactor core, they devised a unique way to eliminate the algae. First the algae were separated from the radioactive water by draining them through a swimming pool filter, then they were pulverized inside a water pump.[72]

Several robots, nicknamed "Rover" and "Louie," have been at work at TMI Unit 2 since 1982 collecting concrete and water samples. Another enormous robot, called "Workhorse," has been specially designed to assist in cleaning out TMI Unit 2's containment building basement. Workhorse's developer, Professor William "Red" Whittaker of Carnegie-Mellon University, declares that this $1 million computer-equipped robot can lift heavy pickup trucks and punch holes through concrete walls.[73]

Engineers have had to learn the condition of TMI Unit 2's reactor fuel as well as the extent of damage suffered by the reactor during the 1979 emergency. Since radiation and contamination levels remained dangerously high, a very small television camera was inserted inside the damaged reactor vessel for remote viewing purposes. Nuclear engineers used various techniques to obtain necessary data about the condition of the reactor and the damaged fuel, including removal and analysis of selected fuel and material samples, internal and external ion chamber measurements, sonar exploration of the core void region, etc.[74]

Examinations disclosed that TMI Unit 2's reactor vessel had survived the 1979 accident intact. Although the reactor vessel head and the upper reactor internals were highly contaminated, both appeared relatively undamaged. However, TMI Unit 2's reactor core was unrecognizable and the reactor fuel had been badly damaged in the course of the TMI accident. The upper five feet of the reactor core had become a huge void, and three feet below this void was a mass of loose rubble. Samples of this rubble suggest that some of the reactor fuel had approached the melting point of uranium oxide during the accident. When TMI Unit 2's reactor core can be completely examined, engineers expect to find a variety of conditions ranging from intact fuel rods to voids and rubble. Beneath the reactor core there remain approximately ten to twenty tons of materials which engineers believe to have once been molten.[75]

Nuclear fuel recovery has been difficult. A variety of vacuum systems, long-handled tools and remote manipulators have been used to extract the fuel, which has been carefully stored in specially designed steel canisters. Pending shipment to the Idaho National Engineering Laboratory for further examination and eventual disposal, these canisters of damaged fuel are stored first under water on racks in TMI Unit 2's fuel handling rooms. Specially constructed casks are used to ship this cargo by rail from Pennsylvania to Department of Energy research facilities in Idaho.[76]

At this writing, complete defueling of the TMI Unit 2 reactor was expected to be completed in late 1987. Other cleanup operations still lie ahead. Future recovery operations will involve collection of remaining fuel particles present in the reactor coolant system and the connected auxiliary systems. The decontamination of the containment and auxiliary buildings remains to be completed. Some eight years after the TMI Unit 2 accident began, the containment building's basement remains inaccessible due to high radiation levels. Complete clean-up and recovery operations at TMI Unit 2 are not expected to be completed before 1989 or later.[77]

As previously mentioned, the cost of the TMI Unit 2 clean-up and recovery program has been estimated at $1 billion. Liability insurance, guaranteed through the Price-Anderson Act, soon proved inadequate to meet these expenses. Additional funding has come from various sources. The Edison Electric Institute (EEI), trade association for America's utilities, has sponsored a plan to raise $150 million from U.S. utilities to assist with the TMI Unit 2 clean-up and recovery program. By January 1984, some thirty-eight corporations had pledged $78.1 million for this endeavor. At a January 1984 meeting, EEI's Board of Directors reaffirmed their commitment to funding this assistance plan for TMI Unit 2.[78] A number of agreements have been effected between a consortium of Japanese utilities and Japanese government representatives and GPU Nuclear and the Department of Energy. Through these agreements, Japanese interests have contributed $7 million in funding for participation in nuclear research pertaining to TMI Unit 2 and access to nuclear power research findings from TMI Unit 2 recovery programs.[79] Finally, cooperative agreements among the Department of Energy, the Nuclear Regulatory Commission and GPU Nuclear have eased the enormous TMI Unit 2 clean-up costs. The DOE and the Idaho National Engineering Laboratory have provided research assistance for various related programs. Radioactive wastes from TMI Unit 2 can be disposed of at U.S. government facilities.[80]

Following prolonged legal battles and considerable public opposition, the TMI Unit 1 reactor was permitted to restart on October 3, 1985, over six and a half years after the Unit 2 emergency began. Since the TMI Unit 1 shutdown, Metropolitan Edison had been spending approximately $180 million annually to purchase necessary replacement electrical power for its customers from other utilities. With TMI Unit 1 again in operation, a Metropolitan Edison vice-president indicated that the company's saving for supplemental electrical power purchases would be passed on to its customers, thereby saving industrial users some $36.5 million and residential users approximately $11 million annually. An individual residential Met-Ed customer, using 500 kilowatts of electricity each month, could expect a $26 dollar reduction in his/her annual electric bill following TMI Unit 1's restart.[81]

As TMI Unit 1 continues in operation, clean-up and recovery operations proceed at TMI Unit 2. Although the TMI Unit 2 accident has brought enormous clean-up and recovery problems, the crisis did produce some positive side effects. The emergency highlighted the need for improved safety operations at all U.S. commercial nuclear power facilities. After the TMI Unit 2 accident, the NRC mandated changes in commercial nuclear power reactor design to increase safety of power plant operations. New training programs have been set up to train commercial nuclear power plant operators in emergency procedures based on the lessons of the TMI Unit 2 accident. More stringent licensing requirement for U.S. commercial nuclear reactors are now in place. A useful summary of U.S. commercial nuclear power plant licensing changes made since the TMI Unit 2 accident may be located in Authorizing of Appropriations for U.S. Nuclear Regulatory Commission for Fiscal Years 1982 and 1983 (MC 82-17346 and CIS 82 H 441-13).

Congress passed the Nuclear Safety Research and Demonstration Act of 1980 (94 STAT 3329) to accelerate research dealing with light water reactor (LWR) safety. This legislation provided funding for a five-year program, directed by the U.S. Department of Energy, and programs to study and recommend improvements in the LWR design were begun in 1981.[82]

Emergency planning for possible disasters, such as a possible serious nuclear power plant accident, received a substantial cash contribution from the General Public Utilities Corporation. By terms of a March 1984 plea-bargaining arrangement negotiated between federal authorities and the General Public Utilities Corporation, the utility agreed to pay a $45,000 fine and $1 million dollars into a local (TMI area) emergency disaster planning fund.[83] These monies are to be used to coordinate efforts for emergency disaster planning in the Harrisburg, Pennsylvania region. It was expected that emergency planning procedures developed for the Harrisburg area could serve as models for similar plans in other areas of the United States.

In conclusion, the Three Mile Island accident has had no one single hero or villain. We have seen how design problems, flawed equipment and human error escalated the TMI Unit 2 accident. But at the same time, during the TMI Unit 2 crisis, there is evidence that many government agencies, departments and representatives of private industry cooperated with the plant operators to bring the crippled reactor successfully to a safe shutdown. Governor Thornburgh acted quickly to set up his own emergency management team to deal with the crisis. In the course of the TMI Unit 2 emergency, Governor Thornburgh and Harold Denton, representing the NRC, kept the public informed with as much detailed information as possible.

Introduction 31

President's Carter's visit to the site gave area residents and the nation added confidence that the TMI Unit 2 crisis was being handled competently, and assurance that any necessary evacuation would be carried out in an orderly manner.

Following the TMI Unit 2 emergency, there has been a continued cooperative effort on the part of the Department of Energy, the Nuclear Regulatory Commission, General Public Utilities Corporation and its subsidiaries, and the U.S. nuclear power industry to work together to resolve the TMI Unit 2 clean-up and recovery problems. Undoubtedly these cooperative efforts took place due to the stimulus of the TMI Unit 2 emergency. We can hope that the lessons of TMI Unit 2 can serve to prevent other serious commercial nuclear power plant accidents from occurring in the United States. Chernobyl's accident has shown the world that the health consequences of a serious nuclear accident can extend for hundreds and even thousands of miles from the site of a nuclear emergency.

Three Mile Island: Notes on the Bibliography

The following annotated bibliography contains Three Mile Island-related citations for selected U.S. government publications, including Congressional hearings and reports accessible from the Monthly Catalog of United States Government Publications and/or the Congressional Information Services' CIS Annual Abstracts/Index to Publications of the United States Congress, covering the years 1972 through 1985. Relevant TMI-related citations are also included for items which appeared in the Commonwealth of Pennsylvania's Checklist of Official Pennsylvania Publication, covering the time period 1979 through 1985. Journal article citations related to Three Mile Island are included which appeared in the Index to U.S. Government Periodicals during the years 1979 through 1985. Selected U.S. government-sponsored research reports and conference papers pertaining to TMI are also included which are available from the National Technical Information Service (NTIS), covering the time period 1972 through 1985.

To facilitate the location of the different types of materials, this TMI-related bibliography is divided into four sections. Within each section, a chronological and alphabetical arrangement of citations has been followed. All citations in the four sections have the following common information:

title, author or principal investigator, publication details, date, pagination, bibliography, and a brief abstract.

Section one lists citations from the Checklist of Official Pennsylvania Publications. Each Pennsylvania document citation includes the appropriate document classification and the words "Pa. State Docs." Selected federal publications from the Monthly Catalog of United States Publications and/or CIS Annual Abstracts/Index to Publications of the United States Congress are listed in Section two and include the Monthly Catalog entry number, and/or the CIS accession number as well as the document classification. The Online Computer Library Center (OCLC) identification number is listed to assist in locating materials.

Section three contains selected citations for journal articles which appeared in the Index to U.S. Government Periodicals. Both the title of the article and the title of the journal and pagination have been included. To assist in locating these articles, the following information is presented: the item number which appears in the Monthly Catalog, the order number from the U.S. Government Periodicals on Microfiche and the OCLC identification number when available. For ordering information, prices, and information on the availability of individual articles, the following address is useful:

    U.S. Government Periodicals on Microfiche
    Index to U.S. Government Periodicals
    c/o Infordata International Inc.
    175 East Delaware Place, Suite 4602
    Chicago, IL   60611

Section four contains selected citations for TMI-related reports, appearing between 1972 and 1985, which are available from the National Technical Information Service. To facilitate location, the NTIS order number and price code is given. The OCLC identification number is listed when available. Section five contains selected citations for TMI-related conference papers which are available from NTIS for the years 1979 through 1985. To facilitate ordering, the NTIS order number and price code are given.

Current prices of NTIS documents can be determined by consulting NTIS's Domestic Price Code Directory, covering the United States, Canada and Mexico. The NTIS Foreign Price

Introduction 33

Directory covers all orders sent outside North America. For additional information, please contact:

> National Technical Information Service
> Sales Desk
> Marketing and Customer Services
> U.S. Department of Commerce
> Springfield, VA 22161
> Phone: 703/487-4650

NTIS orders can be made by telephone, mail, Telex, Telefacsimile, Online Ordering (System Development Corporation and Dialog Information Services), and NTIS's Electronic Ordering Service.

Three Mile Island: Other Information Sources

As the accompanying bibliography demonstrates, the Three Mile Island emergency and its ongoing problems are well represented in government publications and government-sponsored research. In conclusion, mention should be made of the Nuclear Regulatory Commission's collection of technical reports, etc. dealing with all aspects of American nuclear power plant construction, licensing and operation. During its period of operations the Atomic Energy Commission, forerunner of the NRC, collected and issued lists with similar information.

The Atomic Energy Commission issued Title Listing of Civilian Power Reactor Docket Literature in Nuclear Science on a monthly basis, January 1972 through June 1973. The title changed several times: Title Listing of Power Reactor Docket Information was the title from July 1973 through December 1974; Power Reactor Docket Information was the title from January 1975 through December 1978. A number of technical documents related to the Three Mile Island nuclear facility development may be located here.

In January 1979 the NRC began issuing the monthly Title List of Documents Made Publicly Available (NUREG-0540). Each issue contains listings of docketed materials associated with civilian nuclear power plants. A second section contains non-docketed materials which the NRS has received or produced in its capacity as the federal regulatory agency for

America's commercial nuclear power plants. It should be noted that docket numbers and non-docket numbers are assigned for classification of NRC materials and have no connection with U.S. court dockets. Numerous technical materials related to TMI Unit 1 and Unit 2 may be located here.

Following the TMI Unit 2 accident, the NRC issued <u>Title List of Documents of Publicly Available Documents for Three Mile Island Unit 2, Docket 50-230</u> (NUREG-0568), derived from <u>Title List of Documents Made Publicly Available</u> (NUREG-0540). This cumulative list of TMI unit 2 documents extends from 1966 through May 21, 1979. A first supplement extended these listings to October 31, 1979, and a second supplement continued the list through November 15, 1979. For subsequent lists of TMI-related publications, the NRC's monthly <u>Title List of Documents Made Publicly Available</u> (NUREG-0540) should be consulted. It should be noted that selected NRC materials are also listed in the <u>Monthly Catalog</u>. The NUREG designation is given for each document in the Monthly Catalog's description.

Among the various books on nuclear power and the Three Mile Island accident, there are a number which have particular use for researchers and students. A useful history of U.S. nuclear power development is <u>The American Atom: A Documentary History of Nuclear Policies from the Discovery of Fission to the Present: 1939-1984</u>, edited by Robert C. Williams and Philip L. Cantelon. <u>Crisis Contained: The Department of Energy at Three Mile Island</u>, also by Philip L. Cantelon and Robert C. Williams, is a history of the Department of Energy's involvement in the TMI Unit 2 accident. This is a very readable history which will have interest for both beginning and experienced researchers. <u>Three Mile Island Sourcebook: Annotations of a Disaster</u>, by Philip Starr and William Pearson, includes selected local newspaper citations for the TMI Unit 2 accident and for Nuclear Regulatory Commission documents related to the TMI Unit 2 accident published prior to 1981. Finally <u>Three Mile Island Accident: Diagnosis and Prognosis</u>, edited by L. M. Toth and others, is a valuable compilation of technical papers developed from a symposium sponsored by the Division of Nuclear Chemistry and Technology at the 189th meeting of the American Chemical Society, April 28-May 3, 1985, in Miami Beach, Florida. These papers focus on the technical aspects of the TMI Unit 2 accident, the chemical problems of the accident, and the TMI Unit 2

Introduction 35

clean-up and recovery activities during the first six years following the emergency. Environmental concerns related to the TMI Unit 2 accident are excluded from consideration at this time as future meetings are expected to consider the environmental consequences of TMI Unit 2.

This bibliography is designed to assist in locating government publications and government-sponsored research concerned with both TMI Unit 1 and TMI Unit 2. Emphasis is given to the causes of the TMI Unit 2 accident, the TMI Unit 2 accident, the subsequent TMI Unit 2 accident investigations, the ongoing clean-up and recovery operations, and the various economic, social and health related studies. All citations have been arranged alphabetically and chronologically within four general sections in an effort to make the bibliography immediately useful for both researchers and students.

Notes

1. "One Year After Chernobyl, A Tense Tale of Survival," New York Times, 6 April 1987, p. 1.

2. General Public Utilities Corporation. 1986 Annual Report (Parsippany, New Jersey, 1987), pp. 32-33.

3. U.S., Department of Energy, U.S. Commercial Nuclear Power: Historical Perspective, Current Status, and Outlook (Washington, D.C.: Government Printing Office, 1982), p. 1.

4. Ibid., p. 3.

5. Ibid., pp. 4-5.

6. Roger C. Williams and Philip L. Cantelon, eds. The American Atom: A Documentary History of Nuclear Policies from the Discovery of Fission to the Present: 1939-1984 (Philadelphia: University of Pennsylvania Press, 1984), pp. 79-92.

7. U.S., Department of Energy. U.S. Commercial Nuclear Power: Historical Perspective, Current Status, and Outlook, pp. 5-6.

8. Ibid., pp. 5-7.

9. Ibid., pp. 7-9.

10. Ibid., pp. 9-12.

11. U.S., Congress, House, Committee on Science and Technology. Subcommittee on Energy Research and Production of the House Committee on Science and Technology on the Price-Anderson Act, 97th Congress, 1st sess., 1981, pp. 1-5.

12. "Rickover, Nuclear Proponent, Is Dead," New York Times, 9 July 1986, pp. A1, A19.

13. "A Week of Billion-Dollar Headaches for the Proponents of Nuclear Power: Coal's Future Could Rival Its Past," New York Times, 22 January 1984, p. E9.

14. Standard and Poor's Industry Surveys (New York: Standard and Poor's Corporation, 1986), vol. 2, October 1986, p. U22.

15. "TMI Area Residents," Washington Post, 10 April 1979, p. 18.

16. Alan F. Shirk, "The Susquehanna--Its Power-ful Growth," Pennsylvania Forest (Fall 1970), pp. 59-62.

17. Ibid., pp. 62-63.

18. John L. Wise, "Planning and Operating the Nuclear Station," Pennsylvania Forest (Fall 1970), pp. 64-67.

19. Philip L. Cantelon and Robert C. Williams. Crisis Contained: The Department of Energy at Three Mile Island (Carbondale and Edwardsville: Southern Illinois University Press, 1982), pp. 8-9.

20. Ibid., pp. 9-10.

21. TMI Facts and Figures: Three Mile Island (Middletown, PA: Metropolitan Edison Company, 1980), pp. 4-8.

22. Ibid., pp. 2-5.

Introduction   37

23. Philip L. Cantelon and Robert C. Williams. Crisis Contained: The Department of Energy at Three Mile Island, pp. 10-11.

24. TMI Facts and Figures: Three Mile Island, pp. 6-9.

25. Ibid., pp. 7-9.

26. Standard and Poor's Industry Surveys, October 1986, vol. 2, p. U20. Useful diagrams available.

27. Philip L. Cantelon and Robert C. Williams. Crisis Contained: The Department of Energy at Three Mile Island, pp. 8-26.

28. Ibid., p. 15.

29. Ibid., p. 12.

30. W.J. Lanouette, "No Longer Can the NRC Say ... ," Bulletin of the Atomic Scientists, vol. 35, no. 6 (June 1979), pp. 7-8.

31. Philip L. Cantelon and Robert C. Williams. Crisis Contained: The Department of Energy at Three Mile Island, pp. 11-12.

32. Ibid., p. 12-13.

33. Ibid., p. 2.

34. Ibid., pp. 2-3.

35. Ibid., pp. 3-4.

36. Ibid., pp. 5-6.

37. Ibid., p. 7.

38. Ibid., p. 6.

39. Report of the President's Commission on the Accident at Three Mile Island, the Need for Change: The Legacy of TMI (Washington, D.C., 1979), pp. 101-103.

40. Philip L. Cantelon and Robert C. Williams. Crisis Contained: The Department of Energy at Three Mile Island, pp. 40-42.

41. Ibid., pp. 27-48.

42. "Public Alarm Would Not Be So Great If There Was A Single Source of Information," New York Times, 30 March 1979, p. A24.

43. "Governor Thornburgh and Harold Denton Hold Joint News Conference," Harrisburg Patriot, 31 March 1979, P. A1.

44. "Three Mile Island Crisis," New York Times, 31 March 1979, P. A1.

45. Philip L. Cantelon and Robert C. Williams. Crisis Contained: The Department of Energy at Three Mile Island, pp. 67-68.

46. "Hershey Evacuation Center Draws Few Visitors," Reading Eagle, 1 April 1979, p. 3.

47. "Government Offices Reopen in Harrisburg," New York Times, 2 April 1979, p. A1.

48. "Nuclear Regulatory Commission Aide Harold R. Denton Says Crisis at Three Mile Island Is Over; Governor Thornburgh Says It Is Safe for Pregnant Women and Young Children to Return to Their Homes," New York Times, 10 April 1979, p. A1.

49. Robert C. Williams and Philip L. Cantelon, eds. The American Atom: A Documentary History of Nuclear Politics from the Discovery of Fission to the Present: 1939-1984, pp. 312-315.

50. Ibid., p. 315.

51. Report of the Governor's Commission on Three Mile Island (Harrisburg: Commonwealth of Pennsylvania, 1980), Frontispiece, ii.

52. Robert C. Williams and Philip L. Cantelon, eds. The

Introduction 39

<blockquote>American Atom: A Documentary History of Nuclear Politics from the Discovery of Fission to the Present: 1939-1984, pp. 317-318.</blockquote>

53. Ibid., p. 318.

54. Who's Who In America: 1982-1983 (Chicago, IL: Marquis Who's Who, Inc., 1982), 42nd ed., vol. 1, p. 1777.

55. Report of the President's Commission on the Accident at Three Mile Island, The: The Legacy of TMI, pp. 61-62.

56. Ibid., pp. 64-68.

57. Ibid., pp. 70-71.

58. Ibid., pp. 63-64.

59. Ibid., pp. 72-73.

60. Three Mile Island: A Report to the Commissioners and to the Public (Washington, D.C.: Government Printing Office, 1980), vol. 2, pts. 1-3.

61. Nuclear Accident and Recovery at Three Mile Island, A Special Investigation (Washington, D.C.: Government Printing Office, 1980), pp. 23-31.

62. Ibid., pp. 161-217.

63. Report of the Governor's Commission on the Accident at Three Mile Island, pp. 1-2.

64. Ibid., pp. 150-154; pp. 200-204.

65. John C. De Vine, Jr. "The Recovery: A Status Report," in L.M. Roth, ed. ... [et al.], The Three Mile Island Accident: Diagnosis and Prognosis (Washington, D.C.: American Chemical Society, 1986), pp. 268-270.

66. Ibid., pp. 269-270.

67. Governor Dick Thornburgh, "Ten Lessons on Emergency Management From the Accident at Three Mile Island," The Journal: A Publication of the Pennsylvania Emergency Management Agency, (Summer 1986), pp. 8-9.

68. Bryan Burrough and Robert E. Taylor, "Cleaning Up Nuclear Sites Is Inexact Science," Wall Street Journal, 22 May 1986, p. 6.

69. Ibid.

70. Ibid.

71. Ibid.

72. Ibid.

73. Ibid.

74. John C. De Vine, Jr. "The Recovery: A Status Report," pp. 270-274.

75. Ibid., pp. 271-272.

76. General Public Utilities Corporation. 1986 Annual Report, p. 10.

77. Ibid., pp. 10-11.

78. Standard and Poor's Industry Surveys, vol. 2, October 1985, p. U22.

79. Ibid., pp. U22-24.

80. John C. De Vine, Jr. "The Recovery: A Status Report," pp. 269-272.

81. "TMI Restart Important to Economic Development," Reading Eagle, 6 October 1985, p. D1.

82. John N. Wilford. "Safety Regulations Drive Up Expense of Nuclear Reactors," New York Times, 27 February 1984, p. A1.

83. "Metropolitan Edison Company Will Pay $45,000 Fine and Create $1 Million Dollar Fund to Help Pay Emergency Planning Near the Plant," New York Times, 1 March 1984, p. A16.

GLOSSARY

ATOMIC ENERGY ACT--federal legislation, consolidated in 1954, providing for the expanded peaceful uses of nuclear power by American private industry.

ATOMIC ENERGY COMMISSION (AEC)--independent federal commission, superseded by the NRC, which was established in 1946 to oversee peacetime uses of nuclear energy in the United States.

BOILING WATER REACTORS (BWR)--one of the two common types of LWR nuclear reactors used in American nuclear power plants. In the BWR, heat from nuclear fission causes water to boil, turning it to steam which flows directly to the turbine.

DENTON, HAROLD--NRC member sent to TMI site during 1979 TMI Unit 2 emergency to serve as spokesperson for the Nuclear Regulatory Commission.

DEPARTMENT OF ENERGY (DOE)--created by federal legislation in 1974 to oversee U.S. energy resources.

EDISON ELECTRIC INSTITUTE (EEI)--trade association for America's electric utility corporations.

ENVIRONMENTAL PROTECTION AGENCY (EPA)--federal agency with supervisory powers for all activities which affect the quality of U.S. environmental resources.

FERMI, ENRICO--scientist who supervised research leading to the first successful sustained nuclear fission reaction in the United States in 1942.

FRAMPTON, GEORGE, JR.--joint head of Special Inquiry Group,

appointed by Nuclear Regulatory Commission, to investigate the TMI Unit 2 accident.

GENERAL PUBLIC UTILITIES CORPORATION (GPU)--parent corporation for the TMI Nuclear Generating Station.

GENERAL PUBLIC UTILITIES NUCLEAR (GPUN)--Subsidiary created to manage nuclear operations of General Public Utilities Corporation (GPU) and its subsidiaries following the TMI Nuclear Unit II accident.

GOVERNOR'S COMMISSION ON THREE MILE ISLAND--Special Pennsylvania Commission appointed by Governor Dick Thornburgh to investigate the consequences of the TMI Unit 2 accident.

HART, GARY (U.S. Senator)--Chairman of special investigation of the TMI accident made by the Subcommittee on Nuclear Regulation of the Committee on Environment and Public Works of the U.S. Senate following TMI Unit 2 accident in 1979.

JERSEY CENTRAL POWER & LIGHT COMPANY (JCP&L)-- one of three companies that jointly own the TMI nuclear generating station. JCP & L is a subsidiary of General Public Utilities Corporation.

LIGHT WATER REACTOR (LWR)--refers to a general type of nuclear reactor using ordinary water as the medium to transfer heat. There are two types of LWR's: Boiling Water Reactors BWR's) and Pressurized Water Reactors (PWR's).

KEMENY COMMISSION--another designation for the President's Commission on the Accident at Three Mile Island.

KEMENY, JOHN--chairman of special presidential commission which investigated the causes of the TMI Unit 2 accident.

METROPOLITAN EDISON COMPANY--one of the three companies that jointly own the TMI nuclear power generating station. Met-Ed was operating the nuclear facility at the time of the TMI accident and is a subsidiary of the General Public Utilities Corporation.

NUCLEAR REGULATORY COMMISSION (NRC)--U.S. federal

Glossary

commission which superseded the AEC and has general supervisory powers for civilian uses of nuclear power.

NUCLEAR REGULATORY COMMISSION-SPECIAL INQUIRY GROUP--the special task force appointed by the NRC to investigate the TMI Unit 2 accident. Its joint heads were Mitchell Rogovin and George T. Frampton, Jr.

ONLINE COMPUTER LIBRARY CENTER, INC. (OCLC)--a not-for-profit computer library service and research organization which operates an international computer network which member libraries use for acquisition and cataloging of library materials, arranging interlibrary loans, etc.

PENNSYLVANIA ELECTRIC COMPANY (PENN-ELEC)--one of three companies that jointly owned the TMI nuclear generating station at the time of the TMI accident. Penn-Elec is a subsidiary of General Public Utilities Corporation.

POWER DEMONSTRATION REACTOR PROGRAM (PDRP)--program begun in 1955 by joint cooperation of the federal government and private industry, to develop experimental nuclear reactors in the United States.

PRESIDENT'S COMMISSION ON THE ACCIDENT AT THREE MILE ISLAND--special federal commission, appointed by President Jimmy Carter, which investigated the TMI Unit 2 accident and issued its findings in a published report.

PRESSURIZED WATER REACTOR (PWR)--one of the two common types of LWR nuclear power plants. In the PWR, heated primary system water is pressurized to keep it from boiling. This water flows into steam generators where heat transfer causes water in a secondary system to boil, producing steam to drive the turbine.

PRICE-ANDERSON ACT--federal legislation enacted first in 1957, which guarantees public liability in the event of a serious U.S. nuclear power plant accident.

RICKOVER, HYMAN, G. (ADMIRAL)--a strong advocate of expanding the uses of nuclear power; a former director of U.S. Navy's Submarine Propulsion Program.

ROGOVIN, MITCHELL--joint head of a Special Inquiry Group

appointed by the Nuclear Regulatory Commission to investigate the TMI Unit 2 accident.

ROGOVIN REPORT--another name for the published report of TMI Unit 2 accident investigation made by the NRC's Special Inquiry Group.

SCRANTON, WILLIAM W., III--Pennsylvania's Lt.-Governor who served as chairman of the Governor's Commission on Three Mile Island.

THORNBURGH, DICK (GOVERNOR)--Pennsylvania's Governor at the time of the TMI Unit 2 accident, who apointed a special commission to investigate the consequences of this accident.

THREE MILE ISLAND (TMI)--site of TMI Unit 1 and TMI Unit 2 on the Susquehanna River, near Harrisburg, Pennsylvania.

THREE MILE ISLAND UNIT 1--Undamaged nuclear reactor was not affected by 1979 accident at Unit 2 and successfully restarted about six and a half years later.

THREE MILE ISLAND UNIT 2--Scene of most serious nuclear power plant accident in the United States to date occurred here between 28 March 1979 and 9 April 1979.

U.S. CONGRESS--HOUSE COMMITTEE ON INTERIOR AND INSULAR AFFAIRS--Subcommittee on Energy and the Environment--one of two Congressional committees, serving as a successor of the JCAE, which takes responsibility for nuclear power-related matters.

U.S. CONGRESS--JOINT COMMITTEE ON ATOMIC ENERGY (JCAE)--special Congressional committee, created in 1946, to oversee civilian nuclear power matters. Since its dissolution in the 1970s, its former responsibilities are shared by the U.S. Senate's Subcommittee on Nuclear Regulation of the Committee on Environment and Public Works and the U.S. House of Representatives Subcommittee on Energy and Environment of the Committee on Interior and Insular Affairs.

U.S. CONGRESS--SENATE COMMITTEE ON ENVIRONMENT AND PUBLIC WORKS--Subcommittee on Nuclear Regulation-one of two Congressional committees, serving as successor to the JCAE, which takes ongoing responsibility for nuclear power-related matters. Following TMI Unit 2 accident, this committee conducted an investigation of that accident for the U.S. Senate.

PART 1

PENNSYLVANIA GOVERNMENT AND COUNTY PUBLICATIONS*

1. TMI-2 LEGAL EFFECTS
   Legal Aspects of the Three Mile Island Accident: Report of the Legal Subcommittee to the Governor's Commission on Three Mile Island. Harrisburg: The Commission, 1979. 71p. Bibl.
   PYT 5312.2:L496a (Pa. State Docs.) OCLC: 6211031
   Summarizes TMI-2's legal problems, with particular attention to laws and legal proceedings pertaining to the TMI-2 emergency. Environmental planners, government officials, and members of the legal profession will find this study useful.

2. TMI-1 RESTART
   Report in Response to NRC Staff Recommended Requirements for Restart of Three Mile Island Station, Unit 1. Prepared by Met-Ed/GPU. [The Company? 1979?].
   PYT 5312.2:R311 (Pa. State Docs.). OCLC 5525337
   Outlines suggested operating procedures and safety measures adopted by Metropolitan Edison Company at TMI-1 to support its petition to the NRC for permission to restart TMI-1. Report is directed to NRC, but other government officials and local planning groups will find useful information.

3. PENNSYLVANIA NATIONAL GUARD
   Three Mile Island Nuclear Incident, 28 March-5 April 1979: The Pennsylvania National Guard. Annville, Pa.: The Adjutant General, 1979. 86p.
   PMA 23.2:T531 (Pa. State Docs.). OCLC 5229062
   The role of the Pennsylvania National Guard during the TMI-2 accident crisis is described, including the Guard's role in assisting with temporary evacuation of area residents. Emergency planners and government officials will find this report informative.

4. TMI-2 ECONOMIC EFFECTS
   Three Mile Island Socio-economic Impact Study. Prepared by the Governor's Office of Policy and Planning ... Harrisburg: the Office, 1979. 105p.

---

*First line of entry consists of entry number and subject of the publication. Title, other bibliographic information, PA document classification and Online Computer Library Center (OCLC) identification number follow.

PYP 712.2:T531m (Pa. State Docs.). OCLC: 6659050
Study made 6 months after TMI-2 accident evaluates immediate impact of TMI accident upon local residents, area businesses and local industry. Local business and industry representatives will find useful business-related information. See also entry no. 9.

5. TMI-2 HEALTH STUDIES
Health-related Behavioral Impact of the Three Mile Island Nuclear Incident: Report submitted to the TMI Advisory Panel on Health Research Studies of the Pennsylvania Department of Health. Prepared by Peter S. Houts, principal investigator. [Harrisburg]: the Department, 1980. 2 vols. passim. Bibl.
PHE 1.2:H4346 (Pa. State Docs.). OCLC: 6920866
Two-part report describes health-related investigations for TMI area residents. No long-term problems are anticipated, with possible exception of prolonged emotional stress. Health professionals will find useful information. See also entry no. 10.

6. TMI-2 HEALTH STUDIES
Health-related Economic Costs of the Three-Mile Island Accident. Prepared by Teh-wei Hu, principal investigator [et al.]. University Park, Pa.: Pa. State University, 1980. 61p. Bibl.
PHE 194.2:H434c (Pa. State Docs.). OCLC: 7464527
Examines financial costs of immediate health-related services for TMI area residents following the TMI accident. Data were obtained from local surveys of households, physicians and health insurance claim records. Report is directed to health professionals and appropriate government agencies. See also entry no. 18.

7. TMI-2 HEALTH STUDIES
A Layman's Guide to Radiological Health Information and Resources: A Selected Bibliography. Prepared as a staff report to the Governor's Commission on Three Mile Island in Cooperation with the Pennsylvania Department of Health. [Harrisburg, Pa.]: Governor's Commission on Three Mile Island, 1980. 21p. Bibl.
PYT 5312:L427g (Pa. State Docs.). OCLC: 6057493
Describes radiological health materials relevant to TMI-2 emergency, including books, monographs, journal articles, and government documents. Bibliography is directed to the general public and will be useful for researchers, health professionals, and government officials.

8. TMI-2 ACCIDENT REPORTS
Report of the Governor's Commission on Three Mile Island: Presented to the Honorable Dick Thornburgh, Governor, Commonwealth of Pennsylvania. Harrisburg: the Commission, 1980. 219p. Bibl.
PYT 5312.2:R311g (Pa. State Docs.). OCLC: 6449405
Summarizes findings of special Pennsylvania commission which investigated the environmental, health and legal consequences of TMI-2 accident. Report has very useful information for government

officials, environmental planners, health professionals, legal professionals and researchers.

9. TMI-2 ECONOMIC EFFECTS
   The Socio-economic Impacts of the Three Mile Island Accident: Final Report. Prepared by the Governor's Office of Policy and Planning in cooperation with the Department of Agriculture ... Harrisburg: the Office, 1980. 208p.
   PYP 712.2:T531mf (Pa. State Docs.). OCLC: 7052443
   Describes economic effects of TMI accident for area residents, business and local industries in the first year following the emergency. See also entry no. 4. Useful for area businesses and local industry planners.

10. TMI-2 HEALTH STUDIES
    Three Mile Island Health Effects Research Program. Prepared by George K. Tokuhata, Director, Division of Epidemiological Research, Pennsylvania Department of Health. Harrisburg: Pa. Dept. of Health, 1980. 3p.
    PHE 194.2:T531m (Pa. State Docs.). OCLC: 7668588
    Brief description of health studies related to the TMI-2 accident. Report is directed to health professionals. See also entry no. 5.

11. TMI-2, PSYCHOLOGICAL EFFECTS OF ACCIDENT
    Three Mile Island: Mental Health Findings. Prepared by Evelyn Bromet, principal investigator, in collaboration with David Parkinson [et al.]. [Pittsburgh, Pa.: Western Psychiatric Institute and Clinic], 1980. 78p. Bibl.
    PPW 1.2:T531m (Pa. State Docs.). OCLC: 7668342
    Clinical report describes effects of prolonged emotional stress upon TMI area residents. Report, directed to health professionals, indicated need for additional research on the effects of prolonged stress upon individuals.

12. TMI-2 ACCIDENT REPORTS
    Three Mile Island Report. Prepared by Select Committee--TMI, James L. Wright, Jr., chairman ... Pennsylvania General Assembly ... Harrisburg: the Select Committee, 1980. 176 leaves.
    PGA 98.27:CTt531 (Pa. State Docs.). OCLC: 6150459
    Pa. General Assembly report suggests TMI-2 accident was preventable. Since plant operators lacked necessary emergency training, they can not be held solely responsible for the accident's development. Report is directed to Pa. government officials and agencies concerned with commercial nuclear power plant regulation and to members of the Pa. legislature.

13. COUNTY EMERGENCY PLANNING
    Cumberland County, Radiological Emergency Response Plan for Incidents at the Three Mile Island Nuclear Power Station. Prepared by Cumberland County Office of Emergency Preparedness (Pa.). Carlisle, Pa.: The Office, 1981. Looseleaf.

PYE 53.2:R129e:C969 (Pa. State Docs.). OCLC: 8001886
Using looseleaf format for updating, county report outlines proposed nuclear accident emergency preparedness plan, together with recommendations for emergency communications systems. Report is directed to local government officials and to appropriate local agencies responsible for emergency disaster planning.

14. COUNTY EMERGENCY PLANNING
Dauphin County Radiological Emergency Response Plan for Incidents at the Three Mile Island Nuclear Station. Prepared by Dauphin County Emergency Management Agency (Pa.). Harrisburg: The Agency, 1981. Looseleaf.
PYE 53.2:R129e:D242 (Pa. State Docs.). OCLC: 7996088
Looseleaf publication describes county preparedness plan for possible serious nuclear power plant emergencies, with suggested improvements for emergency communications systems. Report is directed to local government officials and to appropriate local agencies responsible for emergency disaster planning.

15. PENNSYLVANIA EMERGENCY PLANNING
Disaster Operations Plan Annex E: Fixed Nuclear Facility Incidents. Prepared by Pennsylvania Emergency Management Agency. Harrisburg: The Agency, 1981. Looseleaf.
PYE 53.2:D611o:981 (Pa. State Docs.). OCLC: 7967740
Plan coordinates emergency accident planning at Pa.'s commercial nuclear power plants among municipal, county, and state agencies and officials. Report is directed to appropriate government agencies and officials, and various local and statewide civil defense groups.

16. TMI-2 HEALTH STUDIES
Fetal and Infant Mortality and Congenital Hypothyroidism Around TMI. Prepared by George K. Tokuhata and Edward Digon. Harrisburg: Pennsylvania Department of Health, 1981. 9 leaves.
PHE 194.2:F419i (Pa. State Docs.). OCLC: 7543955
Investigations indicate no significant changes in infant and fetal mortality rates or increases in reported cases of hypothyroidism for TMI area residents since the TMI-2 accident. Report is directed to health professionals.

17. TMI-2 FINANCIAL EFFECTS
Governor Dick Thornburgh's Proposal to Finance the Cleanup of Three Mile Island. [Harrisburg: The Governor's Office], 1981. 20p.
PGV 1.2:G721pf (Pa. State Docs.). OCLC: 7917302
Formula is suggested to apportion TMI-2 clean-up costs among various federal, state and private sources. Establishment of National Energy Research Institute is recommended to handle TMI-2 financing details and to promote commercial nuclear power safety research. Report is directed to appropriate government agencies and to the public.

Pennsylvania Publications 51

18. TMI-2 HEALTH STUDIES
Health-Related Economic Costs of the Three-Mile Island Accident.
Prepared by Teh-wei Hu, principal investigator, and Kenneth S.
Slaysman. University Park, Pa.: Pa. State University, 1981. 96p.
Bibl.
PHE 194.2:H434cs (Pa. State Docs.). OCLC: 9326764
Final report discusses financial costs of health-related services
for TMI area residents following TMI-2 accident. Report is directed
to health professionals. See also entry no. 6.

19. TMI-2 HEALTH STUDIES
Health Studies in the Three Mile Island Area. Prepared by
George K. Tokuhata, Director of Division of Epidemiological Research,
Pennsylvania Dept. of Health. Harrisburg: Pennsylvania Department of Health, 1981. 11p.
PHE 194.2:H434s (Pa. State Docs.). OCLC: 7669823
Suggests TMI area residents have experienced no serious health
problems related to the TMI-2 accident from low-level radiation exposure. Need for additional epidemiological studies is stressed. Report is directed to health professionals and appropriate government
agencies.

20. TMI-2 HEALTH STUDIES
Impact of TMI Nuclear Accident Upon Pregnancy Outcome, Congenital Hypothyroidism and Infant Mortality. Prepared by George
K. Tokuhata, Director, Division of Epidemiological Research ... ,
Pa. Dept. of Health ... [Harrisburg: Pennsylvania Department of
Health, 1981?]. 14 leaves. Bibl.
PHE 68.2:134t (Pa. State Docs.). OCLC: 8064011
Research indicates no changes, since the TMI-2 accident, in
reported cases of congenital hypothyroidism, fetal mortality, and infant mortality for TMI area residents. Report was subsequently
published as a chapter in Energy, Environment and the Economy,
edited by Shyamel K. Majundar, published by Pa. Academy of
Science in 1981.

21. COUNTY EMERGENCY PLANNING
Lancaster County Radiological Emergency Response Plan for
Incidents at the Three Mile Island Nuclear Station. Lancaster, Pa:
The Agency, 1981. Looseleaf.
PYE 53.2R129e:L245 (Pa. State Docs.). OCLC: 7996585
Describes Lancaster County's emergency planning in the event
of future nuclear power plant accidents. Recommendations for improved emergency communications systems are included. Report is
directed to appropriate government agencies and officials and to
local civil defense groups.

22. COUNTY EMERGENCY PLANNING
Lebanon County Radiological Emergency Response Plan for Incidents at the Three Mile Island Nuclear Station. Prepared by Lebanon County Emergency Management Agency (Pa.). Lebanon, Pa:
The Agency, 1981. Looseleaf.

PYE 53.2:R129e:L441 (Pa. State Docs.). OCLC: 7982951
Outlines county plans in the event of a future commercial nuclear accident, with provision for improved emergency communications systems. Report is directed to approriate government agencies and officials and local civil defense groups.

23. TMI-2 HEALTH STUDIES
Maternal Body Weight, Weight Gain During Pregnancy, Height, and Pregnancy Outcome. Prepared by George K. Tokuhata [et al.]. Harrisburg: Pennsylvania Department of Health, [1981]. 69 leaves. Bibl.
PHE 68.2:M425b (Pa. State Docs.). OCLC: 8070253
Research data indicated no significant health changes were observed when pregnant women living in vicinity of TMI were compared with pregnant women living elsewhere in Pennsylvania, during a research study following the TMI-2 accident. Report is directed to health professionals.

24. TMI AREA POPULATION CHANGES
Mobility of the Population Within 5 Miles of Three Mile Island During the Period from August 1979 through July 1980: Report. Prepared by Marilyn K. Goldhaber and Peter S. Houts, principal investigators, [and] Renee DiSabella. [Harrisburg]: Pennsylvania Department of Health, 1981. 54p.
PHE 194.2:M687p (Pa. State Docs.). OCLC: 7842395
Persons moving from TMI area between August 1979 and July 1980 were surveyed. About 25 per cent of them indicated the TMI accident contributed to their decision to move elsewhere. Report is directed to health professionals and to local area economic planning groups.

25. TMI HEALTH STUDIES
Pregnancy Outcome Around Three Mile Island. Prepared by George K. Takuhata, Director, Division of Epidemiological Research, Pennsylvania Dept. of Health. [Harrisburg]: Pennsylvania Department of Health, 1981. 11 leaves. Bibl.
PHE 194.2:P923o (Pa. State Docs.). OCLC: 7669757
Investigations suggest pregnant women, living in vicinity of TMI, had no immediate accident-related health problems following the TMI-2 emergency. Further research is recommended on long-term effects of low-level radiation and psychological effects of prolonged stress. Report is directed to health professionals.

26. PENNSYLVANIA EMERGENCY PLANNING
Report on Survey of Pa. State Agencies Regarding Activities Intended to Protect Persons & Property from Dangers Assocaited with Nuclear Power Plants. Prepared by Legislative Budget & Finance Committee ... [Harrisburg: The Committee, 1981. 229p.
PGA 1.27:CTB927:R425su (Pa. State Docs.). OCLC: 7950992
Indicates major concerns in emergency planning as: waste products removal, environmental monitoring, rapid public information,

accident communications systems, protective clothing & potassium iodide for endangered personnel, & sufficient funds for accident clean-up. Report is directed to appropriate government agencies and officials and to emergency planning groups.

27. TMI-2 HEALTH STUDIES
Three Mile Island Population Registry. Prepared by Marilyn K. Goldhaber. Harrisburg: ... Pennsylvania Department of Health, 1981. 25 leaves. Bibl.
PHE 194.2:T531pr:Report 1 (Pa. State Docs.). OCLC: 8627787
First of projected series of health surveys for persons living within 5 miles of TMI-2 accident site. Criteria for future studies are described. Report is directed to health professionals and appropriate government departments and agencies.

28. COUNTY EMERGENCY PLANNING
York County Radiological Emergency Response Plan for Incidents at the Three Mile Island Nuclear Station. Prepared by York County Emergency Management Agency (Pa.). York, Pa.: The Agency, 1981. Looseleaf.
PYE 53.2:R129e:Y61 (Pa. State Docs.). OCLC: 8002007
Describes York County's emergency plans in the event of a future serious nuclear power plant emergency at Three Mile Island, emphasizing evacuation procedures and emergency communications systems. Report is directed to appropriate government agencies and officials and local civil defense groups.

29. TMI AREA EMERGENCY EVACUATION
Crisis Evacuation During the Three Mile Island Nuclear Accident: the TMI Population Registry. By Marilyn K. Goldhaber and James E. Lehman ... Harrisburg: Pennsylvania Department of Health, 1983. 42 leaves. Bibl.
PHE 194.2:C932:1983 (Pa. State Docs.). OCLC: 9791943
Summarizes findings concerning temporary exodus of some 144,000 persons from Harrisburg area in the aftermath of TMI-2 accident. Paper is directed to health professionals.

PART 2

U.S. GOVERNMENT PUBLICATIONS:
HEARINGS, REPORTS AND TITLE LISTS*

30. TMI ENVIRONMENTAL IMPACT STATEMENT
Final Environmental Statement Related to the Operation of Three Mile Island Nuclear Station Units 1 and 2, Metropolitan Edison Co., Pennsylvania Electric Co., Jersey Central Power & Light Co., Docket Nos. 50-289 & 50-320. 254 p. Bibl.
Mo. Cat.: 82-14700. Class No.: Y3.N88:10/0552.
OCLC: 5533818
Environment statement, from companies developing the TMI nuclear facility, lists probable impact which construction on the TMI facility is expected to have on local environment. Report was directed to Atomic Energy Commission & contains much information for local environmental planners and government egencies.

31. INDEX, TMI-2 DOCUMENTS
Three Mile Island Nuclear Station: Unit 2, Metropolitan Edison Co., Jersey Central Power & Light Co., Pennsylvania Electric Co. (Operating License Stage). Mickey M. Moore, editor. Oak Ridge, Tenn.: U.S. Energy Research and Develop. Admin...., 1976. 24p.
Mo. Cat.: 76-5011. Class No.: ER1.11:Find-50320-R.
OCLC: 2329719
Fiche nuclear docket list includes TMI-2 related items received by the Atomic Energy Commission between 29 May 1968 and 22 October 1975 from the companies who were developing TMI Nuclear Power Station.

32. [No entry]

---

*First line of each entry consists of entry number and subject of the entry. Title and bibliographic details, Monthly Catalog entry number, U.S. Superintendent of Documents' Classification number, Congressional Information Service (CIS) entry number, and Online Computer Library Center (OCLC) identification number follow.

U.S. Government Hearings, etc. 55

33. TMI-2 SAFETY EVALUATION
Supplement No. 2 to the Safety Evaluation in the Matter of Metropolitan Edison ... : Three Mile Island Nuclear Station, Unit No. 2. Docket Number 50-320. Washington, D.C.: U.S. Nuclear Regulatory Commission, 1978. 75p.
Mo. Cat.: 78-16127. Class No.: Y3.N88:10/0107/suppl.2
OCLC: 3929393
NRC report indicates changes to be made prior to licensing with regard to valve closure in the makeup tank, improvements in the emergency cooling system, fire protection system, etc. Report, directed to Metropolitan Edison Co., etc., is of interest for environmental planners and appropriate government agencies.

34. TMI-2, PRESIDENTIAL COMMISSION APPOINTED
Three Mile Island Powerplant Accident, Presidential Commission for Investigation. Washington, D.C.: U.S. Govt. Print. Off., 1979. 2p.
Mo. Cat: 79-17648.
CIS: 79 PL96-12. OCLC: 5186896
Joint resolution, passed by both the House and the Senate, allocates special temporary powers to a special Presidential Commission to investigate the TMI-2 accident.

35. TMI-2 ACCIDENT REPORTS
Accident at the Three Mile Island Nuclear Powerplant, Part 1. Washington D.C.: U.S. Govt. Print. Off., 1979. 270p.
Mo. Cat.: 79-17508. Class No.: Y4.In8/14:96-8/pt.1.
CIS: 79-H441-26. OCLC: 5223951
Hearings, before a Congressional task force May 9-September 25, 1979, examine causes and development of TMI accident, involving serious damage to the reactor's core and shutdown of the station. Hearings have interest for Congress, government agencies and the public.

36. TMI-2 ACCIDENT REPORTS
Accident at the Three Mile Island Nuclear Powerplant, Part 2. Washington D.C.: U.S. Govt. Print. Off., 1979. 404p. Bibl.
Mo. Cat.: Y4.In8/14:96-8/pt.2.
CIS: 79-H441-26. OCLC: 5223951
Hearings investigate causes of TMI-2 accident, focusing on the accident's development and the response from the parent corporation, the NRC, etc. Hearings have a wide audience--members of Congress, appropriate government agencies, and the public.

37. NUCLEAR POWER PLANT SAFETY
Accident at the Three Mile Island Nuclear Power Plant. Industry's Response to the Accident at TMI. Part 11. Washington, D.C.: U.S. Govt. Print. Off., 1979. 175p. Bibl.
Mo. Cat.: 81-6292. Class No.: Y4.In8/14:96-8/pt. 11.
CIS: 80-H441-33. OCLC: 5223951

56   Three Mile Island

Hearings consider various actions and activities initiated by the
nuclear industry after the TMI-2 accident, to improve reactor design
and the safety of operations in U.S. nuclear power plants. Hearings
are directed to Congress, with interest for the public, government
agencies, etc.

38. TMI-2 EMERGENCY PLANNING
    Civil Defense and the Three Mile Island Nuclear Accident.
Washington, D.C.: U.S. Govt. Print. Off., 1979. 14p.
Mo. Cat.: 80-19346. Class No.: Y4.Ar5/2:N88.
CIS: 79-H202-23. OCLC: 6254030
    Report includes assessment of emergency response system during the TMI accident and an evaluation of emergency planning for severe nuclear power plant accidents since the TMI emergency. Report was prepared for a Congressional committee but it has considerable interest for local and state emergency planning groups and appropriate federal, state and local government officials and agencies.

39. TMI-2 EMERGENCY PLANNING
    Civil Defense Aspects of the Three Mile Island Nuclear Accident.
Washington, D.C.: U.S. Govt. Print. Off., 1979. 246p. Bibl.
Mo. Cat.: 80-9186. Class No.: Y4.Ar5/2a:979-980/11.
CIS: 80-H201-7. OCLC: 5901012
    Hearings deal with protection of civilian population during the TMI-2 accident and planning for future nuclear emergencies utilizing the experiments of the TMI-2 accident. Report was prepared for a Congressional committee, and has interest for local emergency planning groups, appropriate government agencies and officials, etc.

40. NUCLEAR POWER PLANT SAFETY
    Comments on the NRC Safety Research Program Budget. Prepared by U.S. Nuclear Regulatory Commission. Washington, D.C.: U.S. Nuclear Regulatory Commission ... , 1979. 47p.
Mo. Cat.: 79-21310. Class No.: Y3.N88:10/0603.
OCLC: 5300980
    NRC report, prepared to support proposed budget requests for U.S. Congress, includes information on immediate nuclear safety problems, highlighted by the TMI-2 emergency, and recommendations for both immediate and long-term research programs concerning U.S. commercial nuclear reactor design and safety procedures.

41. TMI-2 ACCIDENT RECOVERY ACTIVITIES
    Criticality Analyses of Disrupted Core Models of Three Mile Island Unit 2. Prepared by R.M. Westfall [et al.]. [Washington, D.C.]: U.S. Department of Energy, 1979. 111p. Bibl.
Mo. Cat.: 80-18360. Class No.: E1.28:ORNL/CSD/TM-106.
OCLC: 6492720
    Through use of computer-constructed models, calculations are made to estimate the amount of core damage at TMI-2 following the

U.S. Government Hearings, etc. 57

March 1979 accident. Technical report is directed to nuclear engineers and others interested in preparation for TMI-2 clean-up and decontamination activities.

42. NUCLEAR POWER PLANT EMERGENCY PLANNING
Emergency Planning Around U.S. Nuclear Powerplants. Washington, D.C.: U.S. Govt. Print. Off., 1979. 626p. Bibl..
Mo. Cat.: 80-3622. Class No.: Y4.G74/7:N88/5.
CIS: 79 H401-43. OCLC: 5328928
Hearings examine the effectiveness of emergency planning at U.S. nuclear power plants since the TMI-2 accident, emphasizing local, state and federal planning and changes in the licensing of U.S. nuclear facilities since the TMI-2 accident. Hearings are directed to members of Congress, with interest for appropriate government agencies.

43. TMI-2 ACCIDENT RECOVERY ACTIVITIES
Evaluation of Long-term Post-accident Core Cooling of Three Mile Island Unit 2: NRC Staff Report. Washington, D.C.: Division of Systems Safety, Office of Nuclear Reactor Regulation, U.S. Nuclear Regulatory Commission, 1979. 308p. Bibl.
Mo. Cat.: 79-15819. Class No.: Y3.N88:10/0557.
OCLC: 5083484
NRC report suggests probable condition of reactor core of TMI-2, following March 1979 accident, based on knowledge of the accident's development and estimates of the effects of the reactor core's cooling. Technical report will have particular use for nuclear engineers and scientists.

44. TMI-2 ACCIDENT REPORTS
Generation of Hydrogen During the First Three Hours of the Three Mile Island Accident. Prepared by Randall K. Cole, Jr. Washington, D.C.: U.S. Nuclear Regulatory Commission...., 1979. 39p. Bibl.
Mo. Cat.: 80-5408. Class No.: Y3.N88:25/0913.
OCLC: 5754733
Report suggests approximately 35 per cent of the TMI-2 reactor's core oxidized, producing 350 kg. of hydrogen during early part of accident. This hydrogen was trapped in primary coolant system until blocking valve was reopened. Technical report was prepared for Nuclear Regulatory Commission and has value for nuclear engineers and scientists.

45. NUCLEAR POWER PLANT SAFETY
Inflation in Utilities and Energy. Vol. 4. Washington, D.C.: U.S. Govt. Print. Off., 1979. 211p. Bibl.
Mo. Cat.: 80-9204. Class No.: Y4.B85/3/:In3/v.4.
OCLC: 5886383
Hearing includes information on alternative fuel resources, including safety of nuclear power and significance of TMI-2 accident (pp. 39-53). Hearing is directed to a Congressional Committee, and

has interest for the U.S. nuclear power industry and the general public.

46. TMI-2 ACCIDENT REPORTS
Investigation into the March 28, 1979, Three Mile Island Accident. Prepared by Office of Inspection and Enforcement, U.S. Nuclear Regulatory Commission. Washington, D.C.: The Office, 1979. 650p. Bibl.
Mo. Cat.: 79-17354. Class No.: Y3.N88:10/0600.
OCLC: 6089808
NRC investigation suggests plant operators lacked training to deal with the complexities of the TMI-2 accident and could not be blamed solely for the accident's development. Report was developed for the NRC but has interest for the general public, emergency planners, government agencies and officials, etc.

47. TMI-2 HEALTH EFFECTS
Low-Level Ionizing Radiation. Washington, D.C.: U.S. Govt. Print. Off., 1979. 880p. Bibl.
Mo. Cat.: 80-6852. Class No.: Y4.Sci2:96/41.
CIS: 80-H701-9. OCLC: 5836438
Low-level ionization hearings include DOE report on monitoring during TMI accident, including request for needed research on effects of low-level radiation exposure (pp. 50-104, pp. 121-205). Report has interest for health professionals and the public as well as members of Congress.

48. NUCLEAR POWER PLANT SAFETY
1980 Department of Energy Authorization. Vol. 2 (Nuclear Fission). Washington, D.C.: U.S. Govt. Print. Off., 1979. 630p.
Cat. No.: 80-3698. Class No.: Y4.Sci2:96-14.
CIS: 79-H701-62. OCLC: 5062895
Hearings to review decision to curtail Clinch River (Tenn.) commercial nuclear power plant development, give special attention to similar operational and design problems also present at TMI-2 and Enrico Fermi plant accidents (pp. 553-627). The technical discussion of design problems will be most useful to nuclear engineers and scientists.

49. TMI-2 ENVIRONMENTAL EFFECTS
Non-radiological Consequences to the Aquatic Biota and Fisheries of the Susquehanna River from 1979 Accident at Three Mile Island Nuclear Station. Prepared by C.R. Hickey, Jr. and R.B. Samworth. Washington, D.C.: Division of Site Safety and Environmental Analysis, Office of Nuclear Reactor Regulation, U.S. Nuclear Regulatory Commission, 1979. 97p.
Mo. Cat.: 80-9131. Class No.: Y3.N88:10/0596.
OCLC: 5863324
Radiation releases into the atmosphere did not cause noticeable changes in aquatic biota or fisheries of the Susquehanna River after the TMI accident. No instances of fish disease or changes in fish

U.S. Government Hearings, etc. 59

spawning season are noted. Investigation for the NRC has useful information for environmental planners and area residents.

50. NUCLEAR POWER PLANT SAFETY
Nuclear Powerplant Safety Systems. Washington, D.C.: U.S. Govt. Print. Off., 1979. 1172p. Bibl.
Mo Cat.: 80-5496. Class No.: Y4.Sci2:96/32.
CIS: 80-H701-7. OCLC: 5711632
Hearings examine commercial nuclear safety matters with particular reference to the TMI-2 accident (pp. 6-16, 82-92, 94-116). Report is directed to members of Congress. Environmental planners and appropriate government agencies will find relevant nuclear safety information.

51. NUCLEAR POWER PLANT SAFETY
Nuclear Powerplant Shutdowns--Who Pays? Washington, D.C.: U.S. Print. Off., 1979. 58p. Bibl.
Mo. Cat.: 79-21482. Class No.: Y4.Ec7:N88
CIS: 79-J841-19. OCLC: 5320110
Hearing focuses on economic problems of U.S. nuclear power plant shutdowns due to accidents and safety problems, emphasizing TMI-1 shutdown, consumer rights, and replacement energy costs. Report is directed to a Congressional committee, but the subject of the report makes it useful to government agencies, environmental planners, and researchers.

52. NUCLEAR POWER PLANT SAFETY
NRC Authorizations: Report of Committee on Environment & Public Works, U.S. Senate, Together with Additional Views to Accompany S. 562. Washington, D.C.: U.S. Govt. Print. Off., 1979. 50p.
Mo. Cat.: 79-16046. Class No.: 96-1:S.rpt.176.
CIS: 79-5323-14. OCLC: 5113647
Committee recommendations reflecting majority response to TMI-2 accident stress need for additional emergency planning for any future nuclear emergencies by the NRC, the U.S. nuclear power industry, and appropriate local and state government agencies. Report is directed to a Senate committee but the subject is relevant for local governmental agencies and environmental planners.

53. TMI-2 ACCIDENT REPORTS
NRC Views and Analysis of the Recommendations of the President's Commission on the Accident at Three Mile Island. Prepared by Office of Nuclear Reactor Regulation, U.S. Nuclear Regulatory Commission. Washington, D.C.: The Office, 1979. 51p.
Mo. Cat.: 81-5064. Class No.: Y3.N88.10/0632.
OCLC: 7280952
Report gives NRC's response to major recommendations of the President's Commission on the Accident at Three Mile Island, with particular attention to the proposed restructuring of the NRC to make it an executive agency. Report has a wide audience--government agencies, the U.S. nuclear power industry, and researchers.

54. NUCLEAR POWER PLANT REGULATION
Nuclear Waste and Facility Siting Policy. Washington, D.C.:
U.S. Govt. Print. Off., 1979. 295p. Bibl.
Mo. Cat.: 80-3593. Class No.: Y4.En2:96-38/pt.1.
CIS: 79-S311-73. OCLC: 5652551
Hearings focus on nuclear waste management & licensing of
U.S. commercial facilities, stressing the impact of the TMI-2 accident
upon federal policies and regulations for the U.S. nuclear power industry. Hearings are directed to members of Congress. There is
useful information for environmental planners and appropriate government agencies.

55. ALTERNATIVE ENERGY RESOURCES
Panel Concerning the Long-Range Energy Needs of This and
Other Nations. Washington, D.C.: U.S. Govt. Print. Off., 1979.
36p.
Mo. Cat.: 80-5495. Class No.: Y4.Sci2:96/12.
CIS: 79-H701-53. OCLC: 5245047
Hearing endorses need for continued research of alternative
energy sources, emphasizing the current need for nuclear power as
an energy source for the U.S. even with the dangers of serious nuclear accidents, as shown by the TMI-2 emergency.

56. TMI-2 HEALTH STUDIES
Population Dose and Health Impact of the Accident at the TMI
Nuclear Station: Preliminary Estimates for Period March 28, 1979
Through April 7, 1979. Prepared by Lewis Battist [et al.]. Washington, D.C.: U.S. Nuclear Regulatory Commission, 1979. 114p. Bibl.
Mo. Cat.: 79-21301. Class No.: Y3.N88:10/0558.
OCLC: 5297662
Report estimates no adverse health effects for the residents
living within the vicinity of Three Mile Island due to the TMI-2 accident. Report was prepared by NRC's Incident Response Center
and has particular use for environmental planners, health professionals, the NRC.

57. TMI-2 RADIATION STUDIES
Radiation Measurements Following the Three Mile Island Reactor
Accident. Prepared by Kevin M. Miller [et al.]. New York, N.Y.:
U.S. Department of Energy, 1979. 20p. Bibl.
Mo. Cat.: 80-16395. Class No.: E1.28:EML-357.
OCLC: 6032343
Technical report lists radiation levels present in atmosphere
at various times following the TMI-2 accident. Report was prepared
for the NRC to give accurate information concerning radiation releases
as a result of the TMI-2 accident.

58. NUCLEAR POWER PLANT SAFETY
Radiation Protection. Part 3. Washington, D.C.: U.S. Govt.
Print. Off., 1979. 238p. Bibl.
Mo. Cat.: 80-6803. Class No.: Y4.G74/9:R11/pt.3.
CIS: 80-S401-6. OCLC: 5461747

U.S. Government Hearings, etc. 61

Hearings evaluate radiation monitoring, emergency planning and plant siting at U.S. commercial nuclear power plants, with special attention to the radiation safety issues raised by the TMI-2 accident. Hearings are directed to members of Congress, with useful information for environmental planners, etc.

59. TMI-2 ACCIDENT REPORTS
Report of the Emergency Preparedness and Response Task Force. Prepared by Russell R. Dynes [et al.]. Washington, D.C.: The Commission, 1979. 168p. Bibl.
Mo. Cat.: 80-12631. Class No.: Pr39.8:T41/Em3/2.
OCLC: 6069037
Report recommends installation of additional safety equipment, additional training for plant operators, and improvements in emergency planning at all U.S. commercial nuclear power plants. It is a staff report for the President's Commission on the Accident at Three Mile Island.

60. TMI-2 ACCIDENT REPORTS
Report of the Office of Chief Counsel on Emergency Preparedness, Emergency Response. Prepared by Stanley M. Gorinson [et al.]. Washington, D.C.: The Commission, 1979. 229p. Bibl.
Mo. Cat.: 80-8725. Class No.: Pr39.8:T41/Em3.
OCLC: 9326689
Suggests various safety measures and emergency planning proposals to improve operations at U.S. nuclear power plants in the light of findings concerning the TMI-2 accident. Staff report for the President's Commission on the Accident at Three Mile Island.

61. TMI-2 ACCIDENT REPORTS
Report of the Office of Chief Counsel on the Nuclear Regulatory Commission. Prepared by Stanley M. Gorinson [et al.]. Washington: The Commission, 1979. 139p. Bibl.
Mo. Cat.: 80-8726. Class No.: Pr39.8:T41/N88.
OCLC: 5952078
Report recommends administrative changes in NRC and increased powers to enforce its decisions in the light of TMI-2 accident findings. Staff report for the President's Commission on the Accident at Three Mile Island.

62. TMI-2 ACCIDENT REPORTS
Report of the Office of Chief Counsel on the Role of the Managing Utility and Its Suppliers. Prepared by Stanley M. Gorison [et al.]. Washington, D.C.: The Commission, 1979. 321p. Bibl.
Mo. Cat.: 80-8727. Class No.: Pr39.8:T41/Ut3.
OCLC: 10998360
Report is concerned with TMI-2's control room design, engineering problems, corporate management, emergency planning procedures, and actions of the parent corporation, its subsidiaries, and principal suppliers. Staff report for the President's Commission on the Accident at Three Mile Island.

63. TMI-2 ACCIDENT REPORTS
Report of the President's Commission on the Accident at Three Mile Island: the Need for Change--the Legacy of TMI. Washington, D.C.: The Commission, 1979. 201p. Bibl.
Mo. Cat.: 80-6415. Class No.: Pr39.8.T41/T41.
OCLC: 5814976
Report summarizes findings and recommendations of President's Commission regarding the TMI-2 accident causes. This very significant report draws the widest possible audience, namely the President, the Congress, federal agencies, researchers and the public. It is a "must" for any serious research study of the TMI-2 accident.

64. TMI-2 HEALTH STUDIES
Reports of the Public Health and Safety Task Force on Public Health & Safety Summary, Health Physics and Dosimetry, Radiation Health Effects, Behavioral Effects, Public Health and Epidemiology. [By Jacob I. Fabrikant.] Washington, D.C.: The Commission, 1979. 423p. Bibl.
Mo. Cat.: 80-12632. Class No.: Pr39.8:T41/H34.
OCLC: 6084032
Report describes various proposed health studies for TMI area residents. Need is indicated for long-term low-level radiation studies and for studies of psychological effects of prolonged stress for TMI area residents. Report has immediate interest for health professionals and researchers. Report was developed for the President's Commission on the Accident at TMI.

65. TMI-2 ACCIDENT REPORTS
Report of the Public's Right to Information Task Force. By David M. Rubin [et al.]. Washington, D.C.: The Commission, 1979. 262p. Bibl.
Mo. Cat.: 80-12633. Class No.: Pr39.8:T41/In3.
OCLC: 6082469
Report recommends ways whereby necessary information on a nuclear emergency can be made more readily available to appropriate federal, state, and local agencies and to the public. Report was developed for the President's Commission on the Accident at TMI.

66. TMI-2 ACCIDENT REPORTS
Report of Special Review Group, Office of Inspection and Enforcement, on Lessons Learned from Three Mile Island. Washington, D.C.: U.S. Nuclear Regulatory Commission, 1979. 194p. Bibl.
Mo. Cat.: 80-9133. Class No.: Y3.N88:10/0616.
OCLC: 5882650
NRC report suggests contributing factors to the TMI-2 accident were unsolved problems in reactor safety and control room design, inadequate training in emergency procedures for plant operators, and inadequate emergency communications systems both within the plant and between the NRC and the plant. Report has immediate interest for nuclear engineers, appropriate government agencies, and researchers.

U.S. Government Hearings, etc. 63

67. TMI-2 ACCIDENT REPORTS
Report of the President's Commission on the Three Mile Island Accident. Washington, D.C.: U.S. Govt. Print. Off., 1980. 159p.
Mo. Cat.: 80-15728. Class No.: Y4.D96/10:96-H34.
CIS: 80-S321-14. OCLC: 6267978
Joint hearing of several Congressional committees considers findings of President's Commission on the Accident at TMI and the Commission's recommendations. Testimony by various commission members (pp. 14-59) explain decision not to recommend a moratorium on further nuclear power plant construction and operating licenses, etc.

68. NUCLEAR POWER PLANT SAFETY
Staff Report on the Generic Assessment of Feedwater Transients in Pressurized Water Reactors Designed by Babcock and Wilcox Company. Prepared by U.S. Nuclear Regulatory Commission. Washington, D.C.: U.S. Nuclear Regulatory Commission, 1979. 350p. Bibl.
Mo. Cat.: 79-18864. Class No.: Y3.N88:10/0560.
OCLC: 5003466
NRC technical report discusses the performance of the TMI-2 feedwater transient during the TMI-2 emergency. Findings and recommendations are presented for use by U.S. Commercial nuclear facilities having reactors similar in design to the damaged TMI-2 reactor produced by Babcock and Wilcox.

69. ACCIDENT RECOVERY ACTIVITIES
Safety Evaluation and Environment Assessment: Metropolitan Edison Company, Jersey Central Power and Light Company, Pennsylvania Electric Company. Docket no. 50-320, TMI Nuclear Station, Unit no. 2. Prepared by U.S. Nuclear Regulatory Commission. Washington, D.C.: The Commission, 1980. 19p. Bibl.
Mo. Cat.: 84-15854. Class No.: Y3.N88:19/0647.
OCLC: 10640663
NRC technical report includes NRC requirements with which the involved companies must comply to maintain TMI-2 reactor and reactor building while various clean-up and decontamination activities are taking place.

70. TMI-2 HEALTH STUDIES
Three Mile Island Nuclear Accident, 1979. Washington, D.C.: U.S. Govt. Print. Off., 1979. 143p. Bibl.
Mo. Cat.: 80-9318. Class No.: Y4.L11/4:T41.
CIS: 80-S541-2. OCLC: 5884453
Hearing investigates possible health problems resulting from releases of low-level radiation from TMI-2 accident. Long-term health studies of low-level radiation effects are proposed; no long-term health problems are predicted. Hearing is directed to Congress; appropriate government agencies, and health professionals will find relevant information.

71. TMI-2 ACCIDENT REPORTS
Three Mile Island Nuclear Power Plant Accident. Washington,
D.C.: U.S. Govt. Print. Off., 1979. 254p. Bibl.
Mo. Cat.: 80-13324. Class No.: Y4.Sci2:96/54.12/pt.1.
CIS: 80-H701-15. OCLC: 6093703
Hearing covers events of TMI-2 accident; and the varied actions of the plant operators, parent corporation, NRC & other federal, state and local authorities. Hearing discusses environmental radiation program in TMI vicinity, noting Pennsylvania and local emergency evacuation plans in the event of another severe nuclear power plant emergency.

72. TMI-2 HEALTH STUDIES
Three Mile Island Nuclear Plant Accident. Part 1. Washington,
D.C.: U.S. Govt. Print. Off., 1979. 254p. Bibl.
Mo. Cat.: 80-9336. Class No.: Y4.P96/10:96-H12/pt.1.
CIS: 79-S321-34. OCLC: 5921293
Hearing focuses on possible health or environmental problems related to TMI accident for area residents. Federal agency and department activities at TMI-2 are described and clean-up plans are outlined. Hearing is directed to members of Congress; appropriate government agencies and health professionals will find useful information.

73. TMI-2 ACCIDENT REPORTS
Three Mile Island Nuclear Power Plant Accident. Part 2. Washington, D.C.: U.S. Govt. Print. Off., 1979. 455p. Bibl.
Mo. Cat.: 80-11178. Class No.: Y4.P96/10:96-H12/pt.2.
CIS: 80-S321-14. OCLC: 5921293
Hearings focus on first 48 hours of TMI-2 accident, with evidence derived from personal interviews, transcripts of telephone conversations made to and from the TMI-2 control room and transcripts of discussions of the NRC commissioners. Hearings are directed to members of Congress, with immediate interest for nuclear engineers, emergency planners, appropriate government agencies and researchers.

74. TMI-2 ACCIDENT RECOVERY ACTIVITIES
Three Mile Island Nuclear Powerplant Accident. Part 3: Cleanup and Recovery. Washington, D.C.: U.S. Govt. Print. Off., 1979. 337p. Bibl.
Mo. Cat.: 80-13304. Class No.: Y4.P96/10:96-H12/pt.3.
CIS: 80-S321-16. OCLC: 5921293
Hearings deal with immediate clean-up & recovery activities at TMI-2, made by GPU, Met-Ed, & NRC, following the TMI-2 accident. Huge financial costs and decontamination activities limit clean-up efforts. Hearings are directed to members of Congress, with information of interest to appropriate government agencies, environmental planners, nuclear engineers, and researchers.

75. TMI-2 ACCIDENT RECOVERY ACTIVITIES
Three Mile Island, Unit 2, Radiation Protection Program: Report

U.S. Government Hearings, etc.   65

of the Special Panel. Prepared by C.B. Meinhold [et al.]. Washington
D.C.: The Commission, 1979. 40p.
Mo. Cat.: 80-24699. Class No.: Y3.N88:10/0640.
OCLC: 5863294
   NRC special panel outlines plan designed to safely remove krypton gas and contaminated water from TMI-2 reactor building. Report is directed to The Nuclear Regulatory Commission with interest for environmental planners, engineers, scientists, health professionals, and researchers.

76. TMI-2 HEALTH STUDIES
   Summary and Discussion of Findings from Population Dose and Health Impact of the Accident at the TMI Nuclear Station: (Preliminary Estimate for the Period March 28 through April 17, 1979). Prepared by the Ad Hoc Interagency Population Dose Assessment Group. Rockville, Md.: Bureau of Radiological Health, 1979. 7p.
Mo. Cat.: 79-20401. Class No.: HE20.4102:P81
OCLC: 5327086
   Preliminary estimates indicate no long-term health problems are likely to occur as a result of the March 1979 TMI-2 accident. Research is recommended to evaluate the effects of low-level radiation upon area residents. This HEW report has interest for health professionals, appropriate government agencies, and researchers.

77. TMI-2 FINANCIAL EFFECTS
   Supplemental Appropriations, Communication From the President. Washington, D.C.: U.S. Govt. Print. Off., 1979. 6p.
CIS: 79-H180-84.
   Text of presidential request for additional appropriations from Congress for various nuclear programs, including funds for the NRC to implement recommendations of the Kemeny Commission following the TMI-2 accident. Communication is directed to members of Congress.

78. INDEX, TMI-2 DOCUMENTS
   Title List, Publicly Available Documents, Three Mile Island Unit 2, Docket 50-320. Cumulated to May 21, 1979. Prepared by Division of Technical Information and Document Control, U.S. Nuclear Regulatory Commission. Washington, D.C.: The Division, 1979. 159p. Bibl.
Mo. Cat.: 79-21304. Class No.: Y3. N88:10/0568.
OCLC: 5271376
   List contains TMI-2 materials received by the Atomic Energy Commission and NRC from earliest applications made on April 22, 1968 through May 21, 1979. List is divided into by pre-accident and post-accident documents with category subdivisions. Index has particular interest for nuclear engineers, scientists, researchers, etc.

79. INDEX, TMI-2 DOCUMENTS
   Title List, Publicly Available Documents, Three Mile Island

Unit 2, Docket 50-320: Cumulated to June 30, 1979. Prepared by Division of Technical Information and Document Control, U.S. Nuclear Regulatory Commission. Washington, D.C.: The Division, 1979. 225p. Bibl.
No. Cat.: 79-21305. Class No.: Y 3.N 88:10/0568/rev.1.
OCLC: 5271417

Continues listing of TMI-2 materials received by NRC through June 30, 1979. Listing has particular interst for nuclear engineers, scientists, researchers, etc.

80. INDEX, TMI-2 DOCUMENTS
Title List, Publicly Available Documents, Three Mile Island Unit 2, Docket 50-320: July 1, 1979 to October 31, 1979. Prepared by Division of Technical Information and Document Control, U.S. Nuclear Regulatory Commission. Washington, D.C.: The Division, 1979. 107p. Bibl.
Mo. Cat.: 81-2956. Class No.: Y 3.N 88:10/0568/rev.1/suppl.1.
OCLC: 5985116

List of publicly available documents for TMI-2 received by or developed for the NRC between July 1, 1979 and October 31, 1979.

81. INDEX, TMI-1 DOCUMENTS
Title List, Publicly Available Documents, Three Mile Island Unit 1, Docket 50-289: Cumulated to November 16, 1979. Prepared by Division of Technical Information and Document Control, U.S. Nuclear Regulatory Commission. Washington, D.C.: The Division, 1979. 234p. Bibl.
Mo. Cat.: 81-2958. Class No.: Y 3.N 88:10/0631.
OCLC: 6443500

List of publicly available documents for TMI-1, includes docketed materials and nondocketed informational documents received or developed by the AEC and the NRC to November 16, 1979. Listing has special interest for nuclear engineers, scientists, and researchers.

82. NUCLEAR POWER PLANT SAFETY
TMI-2 Lessons Learned: Task Force Status Report and Shortterm Recommendations. Washington, D.C.: U.S. Nuclear Regulatory Commission, 1979. 97p.
Mo. Cat.: 79-21308. Class No.: Y 3.N 88:10/0578.
OCLC: 5369062

Nuclear safety recommendations are made in light of the TMI-2 accident to improve U.S. nuclear power plant design, licensing, operations, and safety procedures. Report, developed for NRC, has interest for nuclear engineers, environmental planners and appropriate government agencies and officials.

83. NUCLEAR POWER PLANT SAFETY
TMI-2 Lessons Learned: Task Force Final Report. Washington, D.C.: U.S. Nuclear Regulatory Commission, 1979. 47p.
Mo. Cat.: 80-3465. Class No.: Y 3.N 88:10/0585.
OCLC: 5619573

U.S. Government Hearings, etc.   67

Final nuclear safety proposals are made to improve overall safety at U.S. nuclear power plant following study of TMI plant design, licensing, general operations and safety procedures. Report, developed for NRC, has interest for nuclear engineers, environmental planners and appropriate government agencies and officials.

84. TMI-2, PSYCHOLOGICAL EFFECTS OF ACCIDENT
Three Mile Island Telephone Survey: Preliminary Report on Procedures and Findings. Prepared by C. B. Flynn. Washington, D.C.: Division of Safeguards, Fuel Cycle and Environmental Research, Office of Nuclear Regulatory Research, U.S. Nuclear Regulatory Commission. 87p.
Mo. Cat.:   80-3520.   Class No.:   Y 3.N 88:25/1093.
OCLC: 5632505
Telephone survey of TMI-2 area residents sought information on the immediate and ongoing effects of the TMI-2 accident, including attitude of respondents toward TMI-1's possible restart and general safety of other U.S. nuclear facilities. Report, developed for NRC, will have particular interest for psychologists and health professionals.

85. TMI-2 ACCIDENT REPORTS
Analysis of the Three Mile Accident and Alternative Sequences. Prepared by R. O. Wooton [et al.]. Washington, D.C.: The Commission, 1980. 1 vol. passim. Bibl.
Mo. Cat.:   83-14906.   Class No.:   Y 3.N 88:25/1219.
OCLC: 6253464
Computer analysis of TMI-2 accident indicates reactor core was partially uncovered between 1.7 hours and 3.5 hours after the accident began. Core damage could have been decreased by earlier closing of the block valve and continued operation of the high pressure injection system. Technical report is directed to the Nuclear Regulatory Commission, with particular use for nuclear engineers and scientists.

86. TMI-2 RADIATION RELEASES
Answers to Questions About Removing Krypton from the Three Mile Island Unit 2 Reactor Building. Prepared by TMI Program Office. Washington, D.C.: The Program Office, 1980. 19p.
Mo. Cat.:   81-2961.   Class No.:   Y 3.N 88:10/0673.
OCLC: 7091639
Possible problems in removing krypton gas from TMI-2 reactor building are reviewed. Information is presented in question and answer format, particularly useful for beginning researchers. Report was developed and published by NRC and has general public interest.

87. TMI-2 ACCIDENT RECOVERY ACTIVITIES
Answers to Questions Frequently Asked About Cleanup Activities at Three Mile Island. Prepared by TMI Program Office. Washington D.C.: The Program Office, 1980. 39p.

Mo. Cat.: 81-7195. Class No.: Y3.N88:10/0732.
OCLC: 7440498
   Report gives information on various aspects of TMI-2 clean-up activities, including plans for removing decontaminated water and krypton gas from TMI-2 reactor building. Information is presented in question and answer format, which is particularly useful for beginning researchers. NRC report has general public interest.

88. TMI ADVISORY PANEL
   Authorizing Appropriations to the Nuclear Regulatory Commission. Washington, D.C.: U.S. Govt. Print. Off., 1980. 28p.
Mo. Cat.: 80-19148. Class No.: X96-2:H.rp.99o/pt.1.
CIS: 80-H443-15. OCLC: 6388700
   Passage of 1981 NRC appropriations is recommended, including the establishment of a TMI Advisory Panel to advise the NRC on continuing clean-up and decontamination of TMI-2. Report is directed to members of Congress, with relevant information for the NRC, appropriate government agencies, and researchers.

89. NUCLEAR POWER PLANT LICENSING
   Clarification of TMI Action Plan Requirements. Prepared by Division of Licensing, U.S. Nuclear Regulatory Commission. Washington, D.C.: The Commission, 1980. 1 vol. passim. Bibl.
Mo. Cat.: 82-24710. Class No.: Y3.N88:10/0737.
OCLC: 8652682
   Report outlines plan for licensing U.S. commercial nuclear power facilities, stressing need for safe decontamination of TMI-2. NRC report is directed to appropriate government agencies and is useful for researchers, etc.

90. TMI AREA MAP
   Computer-plotted Map of Land Use and Land Cover, Three Mile Island and Vicinity, with Census Tracts. Reston, Va.: U.S. Department of the Interior, Geological Survey, [1980?]. 2p.
Mo. Cat.: 80-18860. Class No.: I19.2:T41/2.
OCLC: 6459064
   Computer-constructed map shows Three Mile Island and adjacent land areas, indicating types of land use and land cover and census tract boundaries. Map has use for environmental planners, appropriate government agencies, geologists, and researchers.

91. TMI-1 RESTART
   Control Room Design Review Report for TMI-1: Metropolitan Edison Company, [et al.], TMI Nuclear Station, Unit no. 1, Docket 50-289. Washington, D.C.: U.S. Nuclear Regulatory Commission, 1980. 41p.
Mo. Cat.: 82-24713. Class No.: Y3.N88:10/0752.
OCLC: 8678107
   Plan outlines specific changes in control room design which must be made before NRC will consider issuing a permit to allow TMI-1 to restart. Technical report has particular information for nuclear engineers and scientists.

U.S. Government Hearings, etc.   69

92. TMI-2 ECONOMIC EFFECTS
    Effects of the Accident at Three Mile Island on Residential
    Property Values and Sales. Prepared by H.B. Gamble and R.H.
    Downing, for U.S. Nuclear Regulatory Commission. Washington,
    D.C.: Division of Safeguards, Fuel Cycle and Environmental Research, Office of Nuclear Regulatory Research, U.S. Nuclear Regulatory Commission. 1980. 135p. Bibl.
    Mo. Cat.: 81-7222.   Class No.:   Y3.N88:25/2063.
    OCLC: 7457867
        Report suggests the TMI-2 accident caused no measurable, permanent change in the costs of residential property within 25-mile area of the damaged TMI-2 nuclear facility. Report has particular interest for environmental planners, local business planning groups, government agencies, and researchers.

93. TMI-1 RESTART
    Emergency Preparedness Evaluation for TMI-1: Metropolitan Edison Company, [et. al.], Three Mile Island Nuclear Station, Unit 1, Docket 50-289. Washington, D.C.: U.S. Nuclear Regulatory Commission, 1980. 36p.
    Mo. Cat.: 82-27066.   Class No.:   Y3.N88:10/0746.
    OCLC: 8789573
        NRC report outlines corrective measures to be implemented before Met-Ed's proposed emergency plan for TMI-1's restart can be approved. Any approved emergency plan must be preceded by a successful emergency drill with appropriate government officials in attendance. Appropriate government agencies, environmental planning groups and researchers will have direct interest in this technical report.

94. NUCLEAR POWER PLANT SAFETY
    Energy and Water Development Appropriations, Fiscal Year 1981. Part 1. Washington, D.C.: U.S. Govt. Print. Off., 1980. 649p.
    Mo. Cat. 80-26646. Class No.:   Y4.Ap6/2:En2/2/981/pt.1.
    CIS: 80-S181-18.4.   OCLC: 6737441
        Budget hearings for FY 81 include information on proposed expansion of NRC programs, involving changes in nuclear safety operations and equipment at U.S. nuclear facilities (pp. 514-525, 580-581). Emergency planners, nuclear engineers and appropriate government agencies will find useful materials concerning power plant safety proposals.

95. NUCLEAR POWER PLANT SAFETY
    Energy and Water Development Appropriations, Fiscal Year 1981.
    Part 2. Washington, D.C.: U.S. Govt. Print. Off., 1980. pp. 639-1484+xi.
    Mo. Cat.: 81-4046.   Class No.:   Y4.Ap6/2:En2/2/981/pt.2.
    CIS: 80-S181-21.3.   OCLC: 6737441
        Hearings receive explanation for DOE FY 81 budget request for additional funding for nuclear power plant safety research, based

on findings of the TMI-2 accident (pp. 841-927). Emergency planners, nuclear engineers and researchers will be interested in safety proposals for U.S. commercial power facilities.

96. NUCLEAR POWER PLANT SAFETY
Energy and Water Development Appropriations for 1981. Part 4. Washington, D.C.: U.S. Govt. Print. Off., 1980. 1240p.
Mo. Cat.: 80-13148. Class No.: Y4.Ap6/1:En2/2/981/pt.4.
CIS: 80-H181-35.1. OCLC: 6081522
Hearings consider DOE's FY 81 budget request, stressing need for changes in U.S. nuclear reactor safety operations as a result of TMI-2 accident investigations (pp. 99-172). Emergency planners, nuclear engineers and researchers will have interest in proposed safety changes for U.S. commercial nuclear power plants, based on TMI-2 accident findings.

97. NUCLEAR POWER PLANT SAFETY
Energy and Water Development Appropriations for 1981. Part 5. Washington, D.C.: U.S. Govt. Print. Off., 1980. 929p.
Mo. Cat.: 80-13148. Class No.: Y4.Ap6/1:En2/2/981/pt.5.
CIS: 80-H181-36.3. OCLC: 6081522
Hearings continue DOE's FY 81 budget request, emphasizing need for additional research programs related to U.S. nuclear power plant safety operations following the TMI-2 accident (pp. 448-534). Hearings are directed to members of Congress, with useful information for nuclear engineers, environmental planners, appropriate government agencies, and researchers.

98. NUCLEAR POWER PLANT LICENSING
Energy and Water Development Appropriations for 1981. Part 12. Washington, D.C.: U.S. Govt. Print. Off., 1980. 365p.
Mo. Cat.: 80-17320. Class No.: Y4.Ap6/1:En2/2/981/pt.12.
CIS: 80-H181-73.3. OCLC: 6081522
Hearings include information on revised NRC requirements for nuclear power plant licensing following TMI-2 accident, together with a recommended timetable for TMI-2 recovery activities (pp. 317-338). Hearings are directed to members of Congress, with relevant information for nuclear engineers, appropriate government agencies and researchers.

99. TMI-2 ENVIRONMENTAL EFFECTS
Environmental Assessment of Radiological Effluents from Data Gathering and Maintenance Operation on Three Mile Island Unit 2. Interim Criteria.... Prepared by TMI Program Office, U.S. Nuclear Regulatory Commission. Washington, D.C.: The Program Office, 1980. 8p.
Mo. Cat.: 81-2962. Class No.: Y3.N88:10/0681.
OCLC: 6680065
Environmental appraisal suggests that probably effects of removal of krypton gas and disposal of decontaminated water in TMI-2 reactor building will not endanger the health of the area population.

U.S. Government Hearings, etc. 71

NRC Report has immediate interest for environmental planners, health professionals, appropriate government agencies, and researchers.

100. TMI-2 RADIATION STUDIES
The Feasibility of Epidemiologic Investigations of the Health Effects of Low-level Ionizing Radiation: Final Report. Prepared by N.A. Dreyer [et al.] for U.S. Nuclear Regulatory Commission. Washington, D.C.: The Commission, 1980. 444p. Bibl.
Mo. Cat.: 81-4028. Class No.: Y3.N88:25/1728.
OCLC: 7216171
Following TMI-2 accident, NRC-sponsored technical study recommends health research projects to determine long-term low-level radiation exposure for nuclear power plant personnel. Health professionals and researchers will find useful information on effects of low-level radiation exposure.

101. TMI-2 HEALTH STUDIES
Federal Radiation Protection Management Act of 1979. Part 2. Washington, D.C.: U.S. Govt. Print. Off., 1980. 231p.
Mo. Cat.: 80-22794. Class No.: Y4.G74/9:R11/pt.2.
CIS: 80-5401-86. OCLC: 6351732
Hearing focuses on need to study health effects of low-level radiation, with particular attention to proposed studies of ionizing radiation for TMI area residents (pp. 129-141). Hearing is directed to members of Congress, with information for health professionals and researchers studying biological and health effects of radiation exposure.

102. TMI-2 ENVIRONMENTAL EFFECTS
Final Environmental Assessment for Decontamination of the Three Mile Island Unit 2 Reactor Building Atmosphere. Final NRC Staff Report. Washington, D.C.: The Commission, 1980. 2 vols. passim. Bibl.
Mo. Cat.: 82-27049. Class No.: Y3.N88:10/0662/v.1-2.
OCLC: 6636781
Report describes the probable consequences of various methods designed to remove the radioactive wastes and to decontaminate TMI-2's reactor building. Alternative plans are evaluated by private groups and governmental agencies or officials. Report is directed to the Nuclear Regulatory Commission; environmental planners, appropriate government agencies, and researchers will also find this report useful.

103. TMI-2 ACCIDENT REPORTS
Fiscal Year 1981 Budget Review. Washington, D.C.: U.S. Govt. Print. Off., 1980. 515p.
Mo. Cat.: 80-19518. Class No.: Y4.P96/10:96-H40.
CIS: 80-S321-24.3. OCLC: 6502967
Hearings, to consider FY 81 budget requests for various federal departments, include statements on U.S. nuclear power plant safety and a detailed review of the TMI accident and follow-up activities,

provided by the NRC's Special Inquiry Group (pp. 105-123). Hearing is directed to members of Congress; researchers and appropriate government agencies may find this summarized information useful.

104. NUCLEAR POWER PLANT SAFETY
H.R. 6910: National Technology Foundation Act of 1980.
Washington, D.C.: U.S. Govt. Print. Off., 1980. 890p. Bibl.
Mo. Cat.: 81-6347. Class No.: Y4.Sci2:96/179.
CIS: 81-H701-49.2. OCLC: 7361254
Hearings, to consider the establishment of a National Technology Foundation (NTF), receive a statement from John Kemeny on the need for additional nuclear safety research programs for U.S. nuclear power plants in the light of the TMI-2 accident findings. The Kemeny statement is reprinted from a journal article, and has interest for the public. Hearings are directed to members of Congress.

105. TMI-2 ACCIDENT REPORTS
Human Factors Evaluation of Control Room Design and Operator Performance at Three Mile Island-2: Final Report. Prepared by T.B. Malone [et al.]. Washington, D.C.: U.S. Govt. Print. Off., 1980. 3 vols. passim. Bibl.
Mo. Cat.: 80-9144. Class No.: Y3.N88:25/1270/v.1-3.
OCLC: 5961603
NRC technical report suggests human errors, made during the first 150 minutes of the accident, were due mainly to flaws in equipment and control room design, and lack of operator training in emergency procedures for a nuclear emergency. Technical report has particular use for nuclear engineers, researchers, etc.

106. TMI-2 HEALTH STUDIES
Investigations of Reported Plant and Animal Health Effects in the Three Mile Island Area. Prepared by Gerald E. Sears [et al.] for U.S. Nuclear Regulatory Commission. Washington, D.C.: Office of Nuclear Reactor Regulation, U.S. Nuclear Regulatory Commission. 40p. Bibl.
Mo. Cat.: 81-7196. Class No.: Y3.N88:10/0738.
OCLC: 7457844
None of reported plant or animal investigations have disclosed health problems which are linked to the TMI-2 accident. NRC-sponsored report has particular interest to health professionals, researchers, etc.

107. TMI-2, REACTION TO ACCIDENT
Impact Abroad of the Accident at the Three Mile Island Nuclear Power Plant: March-September 1979. Prepared by the Congressional Research Service, Library of Congress; [edited by Warren H. Donnelly]. Washington, D.C.: U.S. Govt. Print. Off., 1980. 93p. Bibl.
Mo. Cat.: 80-26693. Class No.: Y4.G74/9:N88/10.
CIS: 80-S402-13. OCLC: 6835431

U.S. Government Hearings, etc.   73

Report analyzes the effects of TMI-2 accident upon nuclear power plant development abroad, particularly in Western European countries and Japan. Report was prepared for use of members of U.S. Congress.

108. TMI-2 ACCIDENT REPORTS
      Oversight, Kemeny Commission Findings. Washington, D.C.:
U.S. Govt. Print. Off., 1980. 132p. Bibl.
Mo. Cat.: 80-15741.   Class No.:   Y4.Sci2:96/74.
CIS: 80-H701-39. OCLC: 6262858
      Hearings consider Kemeny Commission findings, emphasizing need for restructuring of the NRC, improved training for plant operators, improvements in emergency planning and additional safety features on operational equipment. Hearings are directed to House Subcommittee on Energy Research and Production.

109. TMI-2 HEALTH STUDIES
      Known Effects of Low-level Radiation Exposure: Health Implications of the TMI Accident, April 1979. Proceedings of a Conference, April 1979. Edited by P.K. Shrivastava. Bethesda, Md.: U.S. Dept. of Health, Education and Welfare, 1980. 147p. Bibl.
Mo. Cat.: 80-14679. Class No.: HE20.3152:R11.
OCLC: 6160104
      Conference report indicates health problems connected with TMI accident are expected to be minimal. Further research is recommended to access effects of low-level radiation exposure. Report has particular interest for health professionals and researchers.

110. TMI-2 HEALTH STUDIES
      1981 DOE Authorization, Environment Programs. Washington, D.C.: U.S. Govt. Print. Off., 1980. 603p. Bibl.
Mo. Cat.: 80-22892. Class No.: Y4.Sci2:96/107.
CIS: 80-H701-75.2. OCLC: 6583590
      Hearings give information on DOE radiation research studies, including an estimate of radiation dangers for population of TMI area residents from the TMI-2 accident (pp. 77-94). Members of Congress and researchers will find this report most useful.

111. NUCLEAR POWER PLANT SAFETY
      1981 DOE Authorization. Volume 2. Washington, D.C.: U.S. Govt. Print. Off., 1980. 719p.
Mo. Cat.: 80-22897. Class No.: Y4.Sci2:96/114
CIS: 80-H701-85.2. OCLC: 6590125
      Hearings review DOE proposed energy research programs for FY 1981, with special attention to nuclear safety research programs pertaining to the TMI-2 accident (pp. 5-62). Members of Congress will find this material useful.

112. TMI-2 ACCIDENT RECOVERY ACTIVITIES
      NRC Action Plan Developed as a Result of the TMI-2 Accident.
Washington, D.C.: The Commission, 1980. Looseleaf in 2 vols. Bibl.

Mo. Cat.: 81-2960. Class No.: Y3.N88:10/0660/v. 1-2.
OCLC: 6482904
Nuclear Regulatory Commission plan outlines plan to accomplish decontamination and clean-up of TMI Unit 2, utilizing federal, state, company and nuclear industry financial contributions. Members of Congress, researchers, etc. will find this report contains useful information on various financial sources for TMI-2 clean-up funding. See also entry no. 135.

113. NUCLEAR POWER PLANT LICENSING
NRC Oversight: Limitations on Intervenors in Licensing Proceedings. Washington, D.C.: U.S. Govt. Print. Off., 1980. 155p. Bibl.
Mo. Cat.: 81-6260. Class No.: Y4.G74/7:N88/9.
CIS: 81-H401-34. OCLC: 7419586
Hearing discusses NRC nuclear reactor licensing procedure changes made since March 1979, with particular reference to the TMI-2 accident (pp. 10-51), and general safety procedures already in place at U.S. commercial nuclear power plants presently in operation.

114. TMI ACCIDENT RECOVERY ACTIVITIES
NRC Plan for Cleanup Operations at Three Mile Island Unit 2. Prepared by Ronnie Lo and Bernard J. Snyder. Prepared for the Office of Nuclear Reactor Regulation, U.S. Nuclear Regulatory Commission. Washington, D.C.: The Program Office, 1980. 24p.
Mo. Cat.: 81-2964. Class No.: Y3.N88:10/0698.
OCLC: 6770878
Report outlines the NRC's plan to accomplish decontamination and clean-up of TMI Unit 2 and includes various NRC recommendations and DOE's waste disposal proposals.

115. TMI-2 ACCIDENT REPORTS
NRC's Response to the Report of the President's Commission on Three Mile Island. Washington, D.C.: U.S. Govt. Print. Off., 1980. 44p.
Mo. Cat.: 80-19466. Class No.: Y4.In8/4:96-104.
CIS: 80-H501-56. OCLC: 6367987
Hearing reviews the NRC Commissioners' reactions to Kemeny Commission proposals, including proposal to replace NRC with an executive-headed agency, the problems of slowdowns in U.S. nuclear powerplant construction and licensing, etc.

116. TMI-2 ACCIDENT REPORTS
Nuclear Accident and Recovery at Three Mile Island: A Report. Washington, D.C.: U.S. Govt. Print. Off., 1980. 430p. Bibl. 430p. Bibl.
Mo. Cat.: 81-6337. Class No.: Y4.P96/10:96-14.
CIS: 80-S322-12. OCLC: 6689725
Hart Commission presents its evaluation of the TMI-2 accident and its consequences, emphasizing the unsolved health, legal, en-

U.S. Government Hearings, etc. 75

vironmental and financial problems created by the TMI accident. This significant report is the product of a lengthy Congressional investigation into the consequences and ongoing problems created by the TMI-2 accident. Researchers will find this report most usefu.

117. TMI-2 ACCIDENT REPORTS
Nuclear Accident and Recovery at Three Mile Island, Staff Studies. Washington, D.C.: U.S. Govt. Print. Off., 1980. 564p.
Print. Off., 1980. 564p. Bibl.
Mo. Cat.: 81-6338. Class No.: Y4.P96/10:96-14A.
CIS: 80-S3221-16. OCLC: 7024964
Report describes TMI-2 accident development, focusing on activities by NRC and U.S. nuclear power industry to correct operational problems highlighted by the TMI accident. Report was developed for U.S. Senate Subcommittee on Nuclear Regulation of the Committee on Environment and Public Works.

118. NUCLEAR POWER PLANT SAFETY
Nuclear Powerplant Safety After Three Mile Island: Report. [Prepared by Gordon Chapman ... with the assistance of Robert Civiak]. Washington, D.C.: U.S. Govt. Print. Off., 1980. 79p.
Mo. Cat.: 80-19529. Class No.: Y4.Sci2:96/JJ.
CIS: 80-H702-14. OCLC: 6371112
Information on U.S. nuclear power plant operations, focusing on safety of operations since the TMI-2 accident and summarizing testimony presented at Subcommittee on Energy Research and Production hearing.

119. NUCLEAR POWER PLANT SAFETY
Nuclear Regulatory Commission: Fiscal Year 1981 Authorizations. Washington, D.C.: Govt. Print. Off., 1980. 333p.
Mo. Cat.: 80-22865. Class No.: Y4.P96/10:96-H43.
CIS: 80-S321-25. OCLC: 6543829
Hearing, on NRC's FY 81 budget request, contains information on TMI accident recovery programs and on NRC's safety changes for U.S. nuclear commercial power plants (pp. 325-330).

120. TMI-2 ACCIDENT REPORTS
Nuclear Regulatory Commission: The Rogovin Report. Washington, D.C.: U.S. Govt. Print. Off., 1980. 93p.
Mo. Cat.: 80-22784. Class No.: Y4.G74/7:R63.
CIS: 80-H401-57. OCLC: 6606132
Hearings review findings and recommendations of NRC Special Inquiry Group, including changes recommended in operator training, reactor monitoring, nuclear industry reporting system, reactor siting, and evacuation planning for U.S. commercial nuclear power plants following the TMI-2 accident. The NRC Group report is frequently referred to as the "Rogovin Report."

121. NUCLEAR POWER PLANT SAFETY
Nuclear Regulatory Process. Part 3. Washington, D.C.:
U.S. Govt. Print. Off., 1980. 272p. Bibl.
Mo. Cat.: 80-19476. Class No.: Y4.In8/14:96-8/pt.3.
CIS: 80-H441-13. OCLC: 6473559
Hearing considers efforts of U.S. commercial nuclear power plants to improve their safety operations and the NRC's policies and activities pertaining to plant siting and plant licensing in light of the TMI-2 accident (pp. 122-130, 244-268).

122. NUCLEAR POWER PLANT SAFETY
Nuclear Safety Research and Development Act of 1980. Washington, D.C.: U.S. Govt. Print. Off., 1980. 88p.
Mo. Cat.: 81-6238. Class No.: Y4.En2:96-152.
CIS: 81-S311-35.2. OCLC: 7322601
Hearings, concerned with improved nuclear safety research for U.S. commercial nuclear power plants, include DOE report on proposed safe disposal of nuclear wastes at TMI-2 (pp. 57-66).

123. TMI-2 EMERGENCY PLANNING
Nuclear Siting and Licensing Process (Limerick Atomic Power Station, Pa.). Washington, D.C.: U.S. Govt. Print. Off., 1980. 309p. Bibl.
Mo. Cat.: 81-2037. Class No.: Y4.In8/14:96-34.
CIS: 81-H441-2.5. OCLC: 7054587
Hearing includes information on Metropolitan-Edison's actions in developing emergency response planning for the TMI Nuclear Power Plant prior to the March 1979 accident (pp. 225-242).

124. TMI-2 ACCIDENT REPORTS
Nuclear Economics. Part 7. Washington, D.C.: U.S. Govt. Print. Off., 1980. 726p. Bibl.
Mo. Cat.: 80-19477. Class No.: Y4.In8/14:96-8/pt 7.
CIS: 80-H441-24. OCLC: 6497435
Hearings deal with all aspects of the TMI-2 accident, with particular attention to the problems of U.S. nuclear development since the accident, the decontamination and clean-up of TMI-2, and subsequent activities and recommendations of the NRC.

125. NUCLEAR POWER PLANT SAFETY
Plans for Improved Safety of Nuclear Power Plants Following the Three Mile Island Accident. Washington, D.C.: U.S. Govt. Print. Off., 1980. 258p.
Mo. Cat.: 80-17447. Class No.: Y4.Sci2:96/81.
CIS: 80-H701-49. OCLC: 6333872
Hearings evaluate corrective nuclear safety measures taken by government agencies and private industry since TMI-2 accident, including U.S. nuclear industry's establishment of the Institute of Nuclear Power Operations (INPO).

126. TMI-2 RADIATION RELEASES
Post-accident Cleanup of Radioactivity at the Three Mile Island

U.S. Government Hearings, etc. 77

Nuclear Station. Compiled by R.E. Brooksbank, W.J. Armento; contributors, D.O. Campbell [et al.]. Oak Ridge, Tenn.: U.S. Dept. of Energy, 1980. 65p. Bibl.
Mo. Cat.: 80-21133. Class No.: E1.28-ORNL/TM-7081.
OCLC: 6559802
Removel of krypton gas from TMI-2's containment buildings is described, as well as proposed decontamination and clean-up activities at TMI-2.

127. TMI-2 FINANCIAL EFFECTS
Potential Impact of License Default on Cleanup of TMI-2. Prepared by Jack O. Roberts and Jerome Saltman. Washington, D.C.: U.S. Nuclear Regulatory Commission, 1980. 164p. Bibl.
Mo. Cat.: 81-2963. Class No.: Y3.N88:10/0689.
OCLC: 7121391
Report describes effects of potential bankruptcy of GPU corporation, parent company for the TMI Nuclear Generating Station. Enormous TMI-2 clean-up costs will bankrupt the parent corporation unless sufficient financial aid can be obtained from other sources to assist with TMI-2 clean-up and decontamination activities.

128. TMI-2 ACCIDENT REPORTS
Preliminary Calculations Related to the Accident at Three Mile Island. Prepared by J.R. Ireland [et al.] for U.S. Nuclear Regulatory Commission. Washington, D.C.: The Commission, 1980. 135p. Bibl.
Mo. Cat.: 82-29436. Class No.: Y3.N88:25/1353.
OCLC: 8859114
NRC report gives estimates of damage to core of TMI-2's reactor, and estimates the amounts of hydrogen produced during the early hours of the accident by use of a simulated profile showing the first three hours of the TMI-2 accident. Technical report has particular use for nuclear engineers and technical researchers.

129. TMI-2 ACCIDENT REPORTS
Report of the President's Commission on the Three Mile Island Accident. Washington, D.C.: Print. Off., 1980. 159p.
Mo. Cat.: 80-15728. Class No.: Y4.P96/10:96-H34.
CIS: 80-S321-14. OCLC: 6267978
Hearing considers findings and recommendations of the Kemeny Commission regarding the TMI-2 accident. The final recommendations of the Kemeny Commission, including supplemental opinions, are included (pp. 58-154). Hearing is directed to members of Congress. Researchers will find useful material concerning the TMI-2 accident as well as various recommendations and opinions not adopted by the Kemeny Commission in its final report.

130. TMI-2 ACCIDENT REPORTS
Report to the Nuclear Regulatory Commission from the Staff Panel on the Commision's Determination of an Extraordinary Nuclear Occurrence (ENO). Prepared by Office of the Executive Director

for Operations, U.S. Nuclear Regulatory Commission. Washington, D.C.: The Commission, 1980. 1 vol. passim. Bibl.
Mo. Cat.: 82-27045. Class No.: Y3.N88:10/0637.
OCLC: 5984184

NRC report suggests the TMI-2 accident was not "an extraordinary nuclear occurrence," as the accident caused no deaths and did not release substantial amounts of radioactive substances into the atmosphere, or injury to persons and property. Technical report has particular significance for scientists, researchers and nuclear engineers.

131. TMI-2 ACCIDENT REPORTS
Reports of the Technical Assessment Task Force. Washington, D.C.: U.S. Govt. Print Off., 1979-1980. 4 vols. passim. Bibl.
Mo. Cat.: 82-6183. Class No.: Pr39.8:T41/As7/vols.1-4.
OCLC: 5973855

Staff reports, prepared for the Kemeny Commission, give detailed assessment of TMI-2 accident development and its consequences. Information is included on training of TMI plant operators, control room design, availability of emergency communications systems, emergency responses of plant operators, and proposed accident recovery and clean-up procedures. These technical reports present detailed information on all aspects of the TMI-2 accident and are particularly useful for researchers.

132. NRC REORGANIZATION
Reorganization Plan No. 1 of 1980. Washington, D.C.: U.S. Govt. Print. Off., 1980. 406p. Bibl.
Mo. Cat.: 81-3059. Class No.: Y4.G74/9:R29/1980/no.1.
CIS: 80-S401-90. OCLC: 7143750

Congressional hearings consider plans to strengthen NRC's role in nuclear regulatory matters. Excerpts from the Rogovin Report describes the TMI-2 accident development (pp. 267-299) and President's plan designed to strengthen NRC's powers (pp. 214-221).

133. NRC REORGANIZATION
Reorganization Plan No. 1 of 1980. Message from the President. Washington, D.C.: U.S. Govt. Print. Off., 1980. 8p.
CIS: 80-H400-1.

Message outlines suggested changes in NRC management to improve the NRC's supervision of U.S. commercial nuclear power plants, including emergency planning, in the light of the TMI-2 accident.

134. TMI-2 EMERGENCY PLANNING
Response to Committee Report on Emergency Planning and Related Safety Issues. Washington, D.C.: U.S. Govt. Print. Off., 1980. 307p.
Mo. Cat.: 80-17389. Class No.: Y4.G74/7:N88/6.
CIS: 80-H401-24. OCLC: 6331642

Congressional hearing evaluates NRC efforts to improve emer-

U.S. Government Hearings, etc. 79

gency planning procedures and evacuation operations found inadequate during the TMI-2 nuclear power plant accident.

135. TMI-2 ACCIDENT REPORTS
Revision 1 of "NRC Action Plan Developed as a Result of the TMI-2 Accident." (NUREG-0660). Washington, D.C.: U.S. Nuclear Regulatory Commission, 1980. 51p.
Mo. Cat.: 81-2959. Class No.: Y3.N88/10/0660/rev.
OCLC: 6482904
NRC plan to decontaminate and clean-up TMI-2 is updated with revised pages. See also entry no. 112.

136. TMI-2 ECONOMIC EFFECTS
The Social and Economic Effects of the Accident at Three Mile Island: Findings to Date. Prepared by Cynthia B. Flynn and J.A. Chalmers. Washington, D.C.: U.S. Nuclear Regulatory Commission, 1980. 107p. Bibl.
Mo. Cat.: 82-14904. Class No.: Y3.N88:25/1215.
OCLC: 6650571.
NRC report covers economic and social effects of TMI-2 accident for local area residents during six months immediately following TMI accident. Data sources include: published documents, newspaper reports, personal interviews, and household surveys.

137. NUCLEAR POWER PLANT SAFETY
Summary of Public Comments and NRC Staff Analysis Relating to Rule-making on Emergency Planning for Nuclear Power Plants.
Prepared by Division of Siting, Health and Safeguard Standards, U.S. Nuclear Regulatory Commission. Washington, D.C.: The Division, 1980. 99p. Bibl.
Mo. Cat.: 81-6135. Class No.: Y3.N88:10/0684
OCLC: 7353238
NRC report summarizes public comments on plans to improve emergency preparedness procedures in case of future commercial U.S. nuclear power plant accidents.

138. NUCLEAR POWER PLANT SAFETY
Supplemental Appropriations for 1980. Volume 2. Washington, D.C.: U.S. Govt. Print. Off., 1980. 546p.
Mo. Cat.: 80-17318. Class No.: Y4.Ap6/1:Ap6/2/980/pt.2.
CIS: 80-H181-53. OCLC: 6346013
Subcommittee hearings include supplemental budget request from NRC for FY 80 to implement various safety measures and nuclear safety research programs related to TMI-2 accident (pp. 401-422).

139. TMI-2 ACCIDENT RECOVERY ACTIVITIES
Three Mile Island Cleanup and Rehabilitation. Washington, D.C.: U.S. Govt. Print. Off., 1980. 240p. Bibl.
Mo. Cat.: 81-3078. Class No.: Y4.In8/14:96-38.
CIS: 80-H441-36. OCLC: 7091283

Congressional hearing deals with problems involved in effecting safe removal of krypton-85 gas from TMI Unit 2's reactor containment building. Various possible solutions are described and consequences are evaluated.

140. TMI-2 FINANCIAL EFFECTS
Three Mile Island: The Financial Fallout: Report. Prepared by the Comptroller General of the United States. Washington, D.C.: U.S. General Accounting Office, 1980. 92p. Bibl.
Mo. Cat.: 81-00385. Class No.: GA1.13:EMD-80-89.
OCLC: 6974429
Total clean-up and decontamination of TMI-2 is estimated at almost one billion dollars. Various suggested ways of securing the necessary financing are outlined.

141. TMI-2 ACCIDENT REPORTS
Three Mile Island: A Report to the Commissioners and to the Public. Prepared by Special Inquiry Group, U.S. Nuclear Regulatory Commission. Washington, D.C.: The Commission, 1980. 1354p. in 2 vols. Bibl.
Mo. Cat.: 80-154-52. Class No.: Y3.N88:25/1250.
OCLC: 5996296
NRC Special Inquiry Group analyzes development of the TMI-2 accident and makes recommendations for improved nuclear reactor and control room designs, new training for power plant operators, and restructuring of licensing of U.S. nuclear power plants, etc. This report, frequently referred to as "The Rogovin Report," is a "must" for any researcher seriously interested in the TMI-2 accident.

142. NUCLEAR POWER PLANT SAFETY
TMI-Related Requirements for New Operating Licenses. Prepared for U.S. Nuclear Regulatory Commission. Washington, D.C.: The Commission, 1980. 43p. Bibl.
Mo. Cat.: 81-1032. Class No.: Y3.N88:10/0694.
OCLC: 6885758
Following its investigation of the TMI-2 accident, the NRC issues new criteria for commercial nuclear power plant licensing, including new safety features for operations equipment, inprovements in operator training, and changes in plant emergency communications systems. Technical report has particular interest for nuclear engineers and environmental planners.

143. TMI-1 RESTART
TMI-1 Restart: Evaluation of Licensee's Compliance with the Short- and Long-Term Items of Section 11 of the NRC Order Dated August 9, 1979; Met. Ed. Co. et al., TMI Nuclear Station, Unit 1, docket 50-289. Washington, D.C.: U.S. Nuclear Regulatory Commission, 1980. 1 vol. passim. Bibl.
Mo. Cat.: 82-27053. Class No.: Y3.N88:10/0680.
OCLC: 8756532

U.S. Government Hearings, etc. 81

Report reviews efforts by Metropolitan Edison Corporation to bring TMI-1 into conformity with the new nuclear safety standards. Items not completed by Med-Ed are listed. Technical report has particular interest for nuclear engineers and environmental planners.

144. TMI-1 RESTART
TMI-1 Restart: Evaluation of Licensee's Compliance with the Short- and Long-term Items of Section II of the NRC Order Dated August 9, 1979, Met. Ed. Co., et al., TMI Nuclear Station, Unit 1, docket 50-289. Washington, D.C.: U.S. Nuclear Regulatory Commission, 1980. 1 vol. passim.
Mo. Cat.: 82-27054. Class No.: Y3.N88:10/0680/suppl.1.
OCLC: 8709905
Supplementary information is presented by Metropolitan Edison Corporation in an attempt to bring TMI-1 into conformity with revised NRC safety standards. Technical report has particular interest for nuclear engineers.

145. TMI-2 ACCIDENT REPORTS
Three Mile Island: The Most Studied Nuclear Accident in His- History: Summary: Report to the Congress. By the Comptroller General of the United States. Washington, D.C.: U.S. General Accounting Office, 1980. 92p. Bibl.
Mo. Cat.: 81-00483. Class No.: GA1.13:EMD80-109/Sum.
OCLC: 7059819
Report lists the TMI-2 accident investigations and the various proposed health research studies for TMI area residents, authorized or begun in 1979-1980.

146. TMI-2 FINANCIAL EFFECTS
Authorizing Appropriations for the Department of Energy for Fiscal Year 1982: Report (to Accompany H.R. 3505). Washington, D.C.: U.S. Govt. Print. Off., 1981. 29p.
Mo. Cat.: 81-10341. Class No.: Y1.1/8:97-60/pt.1.
CIS: 81-H443-6. OCLC: 7525359
Among DOE proposed programs for FY 1982, proposals are included to continue activities related to TMI recovery from March 1979 accident. Revised cost estimates are included for various proposed activities.

147. TMI-2 ACCIDENT RECOVERY ACTIVITIES
Authorization of Appropriations for U.S. Nuclear Regulatory Commission for Fiscal Years 1982 and 1983. Part 1. Washington, D.C.: U.S. Govt. Print. Off., 1981. 650p. Bibl.
Mo. Cat.: 82-17346. Class No.: Y4.In8/14:97-25/pt.1.
CIS: 82-H441-13. OCLC: 8433909
NRC budget hearings for FY 1982 and 1983 include information on continuing activities related to decontamination and clean-up at TMI-2 (pp. 69-146).

148. TMI-2 PHOTOGRAPHY
Color Photographs of the Three Mile Island Unit 2 Reactor Containment Building. Prepared by Gregory R. Leidam and J. Thomas Horan for U.S. Dept. of Energy. Washington, D.C.: The Three Mile Island Operations Office, U.S. Department of Energy. 1981. [1]
Mo. Cat.: 82-23058. Class No.: E1.28:GEND-006/V.1.
OCLC: 8501304
Photographs, taken with remote sensing equipment, indicate condition of interior of TMI-2 reactor containment building after partial decontamination activities have taken place. These photographs are of immediate use to the persons planning TMI-2 decontamination activities.

149. TMI-1 RESTART
Control Room Design Review Report for TMI-1: Metropolitan Edison Company, et al., TMI Nuclear Station, Unit no. 1, Docket 50-289. Washington, D.C.: U.S. Nuclear Regulatory Commission, 1981. 18p.
Mo. Cat.: 82-24714. Class No.: Y3.N88:10/0752/suppl.1.
OCLC: 8678664
NRC report indicates specific changes in TMI-1's reactor control room design which must be made before a restart of TMI Unit 1 can be approved. Technical report has immediate use for those persons in charge of planning for proposed restart of TMI-1.

150. TMI-1 RESTART
DOE Authorization for Fiscal Year 1982. Volume 2. Washington, D.C.: U.S. Govt. Print. Off., 1981. Numbered 825-1394 pp. in vol. 2. Bibl.
Mo. Cat.: 81-14794. Class No.: Y4.En2/3:97-9.
CIS: 81-H361-14.7. OCLC: 7849247
DOE FY 82 budget request includes summary of information on TMI-1 restart planning and some information on the progress of TMI-2 recovery activities (pp. 1370-1384).

151. TMI-2 ACCIDENT RECOVERY ACTIVITIES
Department of Energy Authorizations for Fiscal Year 1982 and and 1983. Washington, D.C.: U.S. Govt. Print. Off., 1981. 154p. Bibl.
Mo. Cat.: 81-12502. Class No.: Y4.In8/14:97-5.
CIS: 81-H441-20. OCLC: 7662104
Budget hearing for FY 82 includes information on status of TMI-2 accident recovery activities (pp. 36-66). Hearing is concerned with U.S. Dept. of Energy's budget for FY 1982-1983 for various programs, including civilian nuclear waste management activities.

152. TMI-2 ACCIDENT RECOVERY ACTIVITIES
Energy and Water Development Appropriations for Fiscal Year 1982. Part 2. Washington, D.C.: U.S. Govt. Print. Off., 1981. numbered 679-1574 pp. + xiv.

U.S. Government Hearings, etc. 83

Mo. Cat.: 82-24831. Class No.: Y4.Ap6/2:En2/2/982/pt.2.
CIS: 81-5181-36. OCLC: 7844807
Hearings, for FY 82 budget requests concerning proposed NRC appropriations, include an evaluation on TMI-2's decontamination and clean-up activities (pp. 935-1109).

153. TMI-1 RESTART
Energy and Water Development Appropriations for 1982. Part 4. Washington, D.C.: U.S. Govt. Print. Off., 1981. 1152p.
Mo. Cat.: 81-7230. Class No.: Y4.Ap6/1:En2/2/982/pt.4.
CIS: 81-H181-25. OCLC: 7298547
During consideration of NRC budget requests for FY 82, information on reasons for NRC denying restart of TMI-1 is given (pp. 1119-1128).

154. NUCLEAR POWER PLANT SAFETY
Energy and Water Development Appropriations for 1982. Part 8. Washington, D.C.: U.S. Govt. Print. Off., 1981. pp. numbered ii, 1057-1895 + x.
Mo. Cat.: 81-7230. Class No.: Y4.Ap6/1:En2/2/982/pt.8.
CIS: 81-H181-29. OCLC: 7298547
DOE request for FY 82 includes testimony on proposed nuclear safety research programs based on assessment of TMI-2 accident (pp. 1180-1206).

155. TMI-2 ACCIDENT RECOVERY ACTIVITIES
Evaluation of Increased Cesium Loading on Submerged Demineralizer System (SDS) Zeolite Beds. Prepared by DOE-SDS Task Force. Washington, D.C.: U.S. Govt. Print. Off., 1981. 128p. Bibl.
Mo. Cat.: 82-18334. Class No.: E1.68:0012.
OCLC: 8200930
Technical report prepared for the U.S. Dept. of Energy describes procedural adjustments made to improve the Submerged Demineralizer System's (SDS) ability to remove radioactive wastes from the damaged TMI-2 reactor. Scientists and nuclear engineers will have interest in this report.

156. TMI-2 ACCIDENT RECOVERY ACTIVITIES
Final Programmatic Environmental Impact Statement Related to Decontamination & Disposal of Radioactive Wastes Resulting from Mar. 28, 1979 Accident, TMI Nuclear Station, Unit 2. Docket No. 50-320, Met. Ed. Co., JCP & L Co., Penn. El. Co. Washington, D.C.: U.S. Nuclear Regulatory Commission, 1981. 2 vols. passim. Bibl.
Mo. Cat.: 81-5072. Class No.: Y3.N88:10/0683/vol.1 and vol. 2.
CIS: 81-5072. OCLC: 7312177
Technical report outlines alternative proposals to complete clean-up activities relating to TMI-2 and to dispose of radioactive waste materials safely. Report has particular significance for environmental planners and those persons in charge of coordinating TMI-2 recovery activities.

157. TMI-2 FINANCIAL EFFECTS
Financial Implications of the Accident at Three Mile Island.
Washington, D.C.: U.S. Govt. Print. Off., 1981. 924p.
Mo. Cat.: 82-6850. Class No.: Y4.In8/14:97-15.
CIS: 81-H441-27. OCLC: 8123343
Hearings examine revised TMI-2 clean-up cost estimates, power replacement costs and need for additional insurance to cope with any future nuclear emergencies as serious as the TMI-2 emergency.

158. TMI-2 FINANCIAL EFFECTS
Fiscal Year 1982 Budget Review. Washington, D.C.: U.S. Govt. Print. Off., 1981. 878p.
Mo. Cat.: 81-11462. Class No.: Y4.P96/10:97-H3.
CIS: 81-S321-13. OCLC: 7558031
FY 82 budget hearings include review of TMI-2 accident recovery activities, with revised estimates of clean-up costs for TMI-2 and prospects for restart of TMI-1 (pp. 157-264).

159. NUCLEAR POWER PLANT SAFETY
Fiscal Year 1982 Department of Energy Authorization. Volume 1. Washington, D.C.: U.S. Govt. Print. Off., 1981. 1240p.
Mo. Cat.: 81-11469. Class No.: Y4.Sci2/97/2.
CIS: 81-H701-53. OCLC: 7642934
DOE budget request for FY 82 includes proposals for continuation of nuclear safety research programs begun as a consequence of the TMI-2 accident (pp. 153-261). Researchers may find this summary information useful.

160. NUCLEAR POWER PLANT SAFETY
Fiscal Year 1982 Department of Energy Authorization (Nuclear Energy, Magnetic Fusion, Small Hydro, Electric Energy & Energy Storage Systems). Volume 4. Washington, D.C.: U.S. Govt. Print. Off., 1981. 985p.
Mo. Cat.: 82-6305. Class No.: Y4.Sci2:97/2.
CIS: 82-H701-6. OCLC: 7976635
DOE budget hearings for FY 82 include detailed description of nuclear reactor safety research activities at TMI-2 and related investigations to permit safe restart of TMI-1.

161. TMI-2 FINANCIAL EFFECTS
Greater Commitment Needed to Solve Continuing Problems at Three Mile Island: Report to the Congress. Prepared by the Comptroller General of the United States. Washington, D.C.: U.S. Govt. Print. Off., 1981. 116p. Bibl.
Mo. Cat.: 82-00468. Class No.: GA1.13:EMD-81-106.
OCLC: 7920042
Report suggests complete decontamination and accident clean-up activities can be completed at TMI nuclear station, only if financial contributions from federal, state, and private sources are increased. See also entry no. 162.

U.S. Government Hearings, etc. 85

162. TMI-2 FINANCIAL EFFECTS
Greater Commitment Needed to Solve Continuing Problems at Three Mile Island: Summary: Report to the Congress. Prepared by the Comptroller General of the United States. Washington, D.C.: U.S. General Accounting Office, 1981. [13] p.
Mo. Cat.: 82-18699. Class No.: GA1.13:EMD-81-106/sum.
OCLC: 8525227
Summary report suggests unsolved financial costs of TMI-2 clean-up activities required additional funding from various federal, industry, state, and local agencies and organizations. See also entry no. 161.

163. TMI-2 ENVIRONMENTAL EFFECTS
Impact of the 1979 Accident at Three Mile Island Nuclear Station on Recreational Fishing in the Susquehanna River. Prepared by C.R. Hickey. Washington, D.C.: U.S. Nuclear Regulatory Commission, 1981. 37p. Bibl.
Mo. Cat.: 82-27067. Class No.: Y3.N88:10/0754.
OCLC: 8792501
Accident-related effects of TMI-2 accident on fishing was temporary. Poor fish harvest, recorded for five months immediately after the TMI-2 accident, was due in part to anglers' concern for possible ecological or environmental changes in the fish or in the river.

164. TMI-2 ACCIDENT REPORTS
Investigation into Information Flow During the Accident at Three Mile Island. Prepared by U.S. Nuclear Regulatory Commission. Washington, D.C.: The Commission, 1981. 1 vol. passim.
Mo. Cat.: 82-24716. Class No.: Y3.N88:10/0760.
OCLC: 7994142
NRC report evaluates the various public communication sources available at the time of the TMI-2 accident. Importance of radio coverage for TMI area residents is noted. The need for improved communications systems during any future nuclear emergency is stressed.

165. TMI-2 LEGAL EFFECTS
Nuclear Powerplant Licensing Delays and the Impact of the Sholly Versus NRC Decision. Washington, D.C.: U.S. Govt. Print. Off., 1981. 310p. Bibl.
Mo. Cat.: 81-14933. Class No.: Y4.P96/10:97-H11.
CIS: 81-S321-18. OCLC: 7833598
Reasons are listed for nuclear power plant licensing delays since TMI-2 accident, emphasizing legal consequences of Sholly vs. NRC decision. Ruling requires NRC to hold public hearings on demand before amending a U.S. commercial nuclear power plant's license.

166. TMI-2 HYDROGEN BURN
Proceedings of the Workshop on the Impact of Hydrogen on

Water Reactor Safety. Vol. 4. Edited by Marshall Berman: [prepared for U.S. Nuclear Regulatory Commission]. Washington, D.C.: The Commission, 1981. 288p. Bibl.
Mo. Cat.: 84-4070. Class No.: Y3.N88:35/2017/v.4.
OCLC: 10052227

Workshop papers, compiled in four volumes totaling 1147 pages, focus on the general topic of hydrogen behavior during a nuclear reactor accident, with an emphasis on reactor safety and related research. A special workshop session, reported in volume 4, discusses details of TMI-2 hydrogen burn, and includes information from authorized entries into TMI-2's reactor building since the TMI-2 emergency ended.

167. TMI-2 HEALTH STUDIES
The Public Whole Body Counting System Following the Three Mile Island Accident. Prepared by R.L. Gotchy and R.J. Bores. Washington, D.C.: U.S. Nuclear Regulatory Commission, 1981. 61p. Bibl.
Mo. Cat.: 82-27062. Class No.: Y3.N88/10/0636.
OCLC: 7620643

Some 762 area residents who lived in the vicinity of TMI-2 during the TMI-2 accident were tested subsequently for unusual radiation accumulations in their bodies. No persons tested were found to have abnormal amounts of radiation.

168. NUCLEAR POWER PLANT SAFETY
Radiological Emergency Planning and Preparedness. Washington, D.C.: U.S. Govt. Print. Off., 1981. 220p. Bibl.
Mo. Cat.: 81-14934. Class No.: Y4.96/10:97-H13.
CIS: 81-S321-19. OCLC: 7831212

Hearings focus on federal, state, and local efforts to coordinate joint emergency planning in the event of future serious nuclear accidents similar to TMI-2 emergency (pp. 3-34).

169. NUCLEAR POWER PLANT SAFETY
Report to Congress on Status of Emergency Response Planning for Nuclear Power Plants. Washington, D.C.: U.S. Nuclear Regulatory Commission, 1981. 1 vol. passim. Bibl.
Mo. Cat.: 82-27068. Class No.: Y3.N88:10/0755.
OCLC: 8792785

In light of TMI-2 accident, NRC report describes emergency response planning, potential communications systems in an emergency, and possible evacuation time estimates for populated areas near U.S. commercial nuclear power plants presently in operation.

170. TMI-2 ACCIDENT REPORTS
Reporting of Information Concerning the Accident at Three Mile Island: A Report. Prepared by the Majority Staff of the Committee on Interior and Insular Affairs.... Washington, D.C.: U.S. Govt. Print. Off., 1981. 158p.
Mo. Cat.: 81-10468. Class No.: Y4.In8/14:Ac2/12.

U.S. Government Hearings, etc. 87

CIS: 81-H442-7. OCLC: 7543717
Staff report examines problems caused by TMI-2 plant operators' failure to transmit timely and adequate information on TMI's accident development to appropriate governmental agencies and officials.

171. NUCLEAR POWER PLANT SAFETY
The Research Program of the Nuclear Regulatory Commission. Washington, D.C.: U.S. Govt. Print. Off., 1981. 69p.
Mo. Cat.: 81-14935. Class No.: Y4.P96/10:97-H16.
CIS: 81-S21-20. OCLC: 7844838
Proposed nuclear safety programs for the NRC FY 82 and FY 83 are outlined, with recommendations in light of the TMI-2 accident findings to improve safety and operational procedures at U.S. commercial nuclear power facilities.

172. TMI-2 ACCIDENT RECOVERY ACTIVITIES
Safety Evaluation Report: Operation of the Submerged Demineralizer System at TMI Nuclear Station, Unit no. 2. Docket no. 50-320: Met. Ed. Co., JCP and L Co., PE Co. Washington, D.C.: U.S. Nuclear Regulatory Commission, 1981. 82p. Bibl.
Mo. Cat.: 81-11318. Class No.: Y3.N88:10/0796.
OCLC: 7632856
Technical report summarized NRC response to Metropolitan Edison and others' proposed operation of the TMI-2 Submerged Demineralizer (SDS), an underwater radioactive waste processing system for TMI-2 radioactive wastes.

173. TMI-1 RESTART
TMI-1 Restart: Evaluation of Licensee's Compliance with Short- and Long-term Items of Section II of NRC Order Dated Aug. 9, 1979. Met. Ed. Co., et al., TMI Nuclear Station, Unit 1, Docket 50-289. Washington, D.C.: U.S. Nuclear Regulatory Commission, 1981. 16p.
Mo. Cat.: 82-27055. Class No.: Y3.N88:10/0680/suppl.2.
OCLC: 8757414
NRC report gives an evaluation of efforts by General Public Utilities Corporation and its subsidiaries to receive a favorable acceptance of plan for restart of TMI-1. Technical report has immediate use for companies interested in restart of TMI-1.

174. TMI-1 RESTART
TMI-1 Restart: Evaluation of Licensee's Compliance with the Short- and Long-term Items of Section II of NRC Order Dated Aug. 9, 1979. Met. Ed. Co., et al., TMI Nuclear Station, Unit 1 Docket 50-289. Washington, D.C.: U.S. Nuclear Regulatory Commission, 1981. 1 vol. passim.
Mo. Cat.: 82-27056. Class No.: Y3.N88:10/0680/suppl.3.
OCLC: 8756473
NRC report continues an evaluation of Metropolitan Edison's efforts to gain permission for restart of TMI-1. Technical report has immediate use for companies interested in restart of TMI-1.

175. TMI-2 RADIATION RELEASES
 Use of Film to Estimate Exposure Near the Three Mile Nuclear Power Station. Prepared by Ralph E. Shuping. Rockville, Md.: U.S. Dept. of Health & Human Services, 1981. 20p. Bibl.
Mo. Cat.: 81-10014. Class No.: HE20.4102:P56/7.
OCLC: 7519307
 Photography of atmosphere during the TMI-2 accident, and for some successive days, suggests the TMI area residents would experience no long-term adverse radiation effects as a result of radiation releases during and following the TMI-2 accident. Report was prepared for use of U.S. Dept. of Health and Human Services.

176. TMI-2 ACCIDENT RECOVERY ACTIVITIES
 Cleanup Efforts at Three Mile Island. Washington, D.C.: U.S. Govt. Print. Off., 1982. 75p.
Mo. Cat.: 82-22416. Class No.: Y4.En2/3:97-108.
CIS: 82-H361-67. OCLC: 8638025
 Congressional hearing examines TMI-2 accident liability responsibilities of the federal government and of the GPU Corporation, stressing the corporation's need for additional financial assistance to complete removal of radioactive wastes and related clean-up activities at TMI-2.

177. TMI-2 FINANCIAL EFFECTS
 Electric Utility Nuclear Accident Cost Allocation Act. Washington, D.C.: U.S. Govt. Print. Off., 1982. 62p.
Mo. Cat.: 83-5857. Class No.: Y1.1/5:97-524.
CIS: 82-S323-16. OCLC: 9013916
 Technical report gives various opinions on proposed plan to establish a supplemental nuclear insurance trust fund, financed by U.S. electric utilities operating nuclear powerplants, to finance remaining clean-up activities at TMI-2.

178. NUCLEAR POWER PLANT SAFETY
 Evaluation of the Prompt Alert Systems at Four Nuclear Stations. Prepared by D.A. Towers [et al.] for U.S. Nuclear Regulatory Commission. Washington, D.C.: The Commission, 1982. 271p.
Mo. Cat.: 83-16419. Class No.: Y3.N88:25/2655.
OCLC: 9502847
 Technical report describes emergency warning systems in place at four U.S. commercial nuclear power plants: Indian Point, Three Mile Island-1, Trojan and Zion facilities.

179. TMI-2 FINANCIAL EFFECTS
 Financial Fallout from Three Mile Island. Washington, D.C.: U.S. Govt. Print. Off., 1982. 392p. Bibl.
Mo. Cat.: 83-6217. Class No.: Y4.En2/3:97-163.
CIS: 82-H361-117. OCLC: 9072497
 Hearings include information on revised TMI-2 clean-up costs, power replacement charges and proposed increased nuclear insurance requirements for all U.S. commercial nuclear power plants.

U.S. Government Hearings, etc. 89

180. TMI-2 FINANCIAL EFFECTS
Financing the Cleanup of the Three Mile Island Nuclear Powerplant. Washington, D.C.: U.S. Govt. Print. Off., 1982. 436p. Bibl.
Mo. Cat.: 82-12141. Class No.: Y4.En2:97-52.
CIS: 82-S311-17. OCLC: 8271543
Joint hearing evaluates projected TMI-2 clean-up costs and future nuclear power plant insurance costs in the event of another serious commercial nuclear power plant emergency.

181. TMI-2 ACCIDENT RECOVERY ACTIVITIES
Fiscal Year 1983 Budget Review. Washington, D.C.: U.S. Govt. Print. Off., 1982. 644p.
Mo. Cat.: 82-24931. Class No.: Y4.P96/10:97-H37.
CIS: 82-S321-27. OCLC: 8664892
Budget hearing includes information on TMI-2 clean-up progress, and includes a report of NRC emergency planning procedures in the event of a future serious commercial nuclear power plant emergency.

182. NUCLEAR POWER PLANT SAFETY
Fiscal Year 1983 Department of Energy Budget Review (Nuclear Fission R. & D. Small-Scale Hydro, Electric Energy Systems & Energy Storage Systems). Volume 4. Washington, D.C.: U.S. Govt. Print. Off., 1982. 596p. Bibl.
Mo. Cat.: 83-10538. Class No.: Y4.Sci2:97-113.
CIS: 82-H701-85. OCLC: 9089071
Hearings include proposed research and development costs related to TMI-2 safety-connected features and safety requirements for all U.S. commercial nuclear power plants (pp. 206-227).

183. NUCLEAR POWER PLANT SAFETY
Health Physics Appraisal Program. Prepared by L.J. Cunningham, et al. for U.S. Nuclear Regulatory Commission. Washington, D.C.: The Commission, 1982. 105p. Bibl.
Mo. Cat.: 82-29376. Class No.: Y3.N88:10/0855.
OCLC: 8863869
Technical report discusses the U.S. Nuclear Regulatory Commission's radiation program, giving particular attention to the need for improved radiation protection at all U.S. commercial nuclear power facilities as evidenced by TMI-2 accident findings.

184. TMI-2 ACCIDENT REPORTS
H.R. 3839: Establish a U.S. Design Council Within the Department of Commerce. Washington, D.C.: U.S. Govt. Print. Off., 1982. 120p.
Mo. Cat.: 83-4216. Class No.: Y4.Sci2:97/92.
CIS: 82-H701-62. OCLC: 8937835
Hearing includes transcripts of article by John G. Kemeny focusing on TMI-2 accident and relating TMI accident causes to a general decline in U.S. industrial design (pp. 74-85).

185. TMI-2 ACCIDENT RECOVERY ACTIVITIES
Impact of Federal R & D Funding on Three Mile Island Cleanup Costs: Report. Prepared by U.S. General Accounting Office.
Washington, D.C.: The Office, 1982. 23p.
Mo. Cat.: 84-8429. Class No.: GA1.13:EMD-82-28.
OCLC: 8351350
Report indicates federal funding appropriated for research pertaining to U.S. nuclear power plant emergency planning will not be used to defray any of the financial costs of TMI-2 clean-up. Additional funding for TMI-2 clean-up activities is recommended.

186. NUCLEAR POWER PLANT SAFETY
Nuclear Fuel Cycle Policy and the Future of Nuclear Power.
Washington, D.C.: U.S. Govt. Print. Off., 1982. 774p.
Mo. Cat.: 82-15149. Class No.: Y4.In8/14:97-18.
CIS: 82-H441-10. OCLC: 8293056
Congressional hearing includes report on anticipated effects of various federal government policies on U.S. nuclear power plant safety and operations, with special attention to the TMI-2 accident findings and safety measures adopted at U.S. commercial power plants since those findings were published (pp. 509-549).

187. TMI-2 FINANCIAL EFFECTS
Nuclear Property Insurance Act of 1981. Washington, D.C.:
U.S. Govt. Print. Off., 1982. 369p. Bibl.
Mo. Cat.: 82-29813. Class No.: Y4.P96/10:97-H52.
CIS: 82-S321-33. OCLC: 8835773
Hearings include information used to determine equitable costs for TMI-2 clean-up with data on GPU Corporation's financial problems and increases in local electric utility rates for consumers (pp. 196-201, 233-269).

188. TMI-2 ACCIDENT RECOVERY ACTIVITIES
NRC Budget Request for Fiscal Year 1983. Washington, D.C.:
U.S. Govt. Print. Off., 1982. 272p.
Mo. Cat.: 82-27252. Class No.: Y4.En2/3:97-132.
CIS: 82-H361-95. OCLC: 8818790
NRC budget hearing includes information on progress of TMI-2 clean-up and a revised estimate of remaining costs (pp. 120-247).

189. TMI-2 ACCIDENT RECOVERY ACTIVITIES
NRC Plan for Cleanup Operations at Three Mile Island Unit 2.
Rev. 1. Washington, D.C.: U.S. Nuclear Regulatory Commission, 1982. 1 vol. passim. Bibl.
Mo. Cat.: 82-27057. Class No.: Y3.N88:10/0698/rev.1.
OCLC: 8802655.
Report gives revisions for TMI-2 clean-up plan previously presented by NRC. Technical report has particular interest for those persons in charge of planning TMI-2 decontamination activities.

U.S. Government Hearings, etc. 91

190. NUCLEAR POWER PLANT SAFETY
Nuclear Safety--Three Years After Three Mile Island. Washington, D.C.: U.S. Govt. Print. Off., 1982. 63p.
Mo. Cat.: 82-24901. Class No.: Y4.G74/7:N88/12.
OCLC: 8653543
Joint hearings before several House subcommittees suggest that new safety features for nuclear power plant equipment and changes in plant operators' training have made power plant operations safer than before the TMI-2 accident.

191. NUCLEAR POWER PLANT SAFETY
Problems and Delays Overshadow NRC's Initial Success in Improving Reactor Operators' Capabilities: A Report. Prepared by U.S. General Accounting Office. Washington, D.C.: The Office, 1982. 43p. Bibl.
Mo. Cat.: 83-17293. Class No.: GA1.13:RCED-83-4.
OCLC: 9361516
Following the TMI-2 emergency, the NRC devised new operational guidelines designed to improve both the quality of operations and emergency preparedness training for reactor operators at U.S. commercial nuclear power facilities. This report discusses areas of progress and also notes continuing problems.

192. TMI-1 RESTART
Proceedings of Workshop on Psychological Stress Associated with the Proposed Restart of Three Mile Island, Unit 1 (1982: McLean, Va.). Edited by Pamela Walker [et al.]. Washington, D.C.: The Commission, 1982. 153p. Bibl.
Mo. Cat.: 83-12388. Class No.: Y3.N88:27/0026.
OCLC: 8932226
Focus of workshop reports is on the effects of prolonged emotional stress upon TMI area residents occasioned by plans for restart of TMI, Unit 1.

193. TMI-2, ECONOMIC AND SOCIAL EFFECTS
Three Mile Island: A Case Study. Volume 12 in Socioeconomic Impacts of Nuclear Generating Stations. Prepared by J. Flynn for U.S. Nuclear Regulatory Commission. Washington, D.C.: The Commission, 1982. 241p. Bibl.
Mo. Cat.: 83-10333. Class No.: Y3.N88:25/2749/v.12.
OCLC: 9223418
Part of a twelve-volume study dealing with U.S. commercial nuclear power facilities, this volume stresses the consequences of the development of the TMI facility for Pennsylvania and, in particular, the immediate area adjacent to this nuclear power facility. Emphasis is placed on the immediate consequences of the TMI-2 accident, 1979-1981.

194. TMI-2 ENVIRONMENTAL EFFECTS
Answers to Questions About Updated Estimates of Occupational

Radiation Doses at Three Mile Island. Unit 2. Washington, D.C.:
TMI Program Office, 1982. 24p.
Mo. Cat.: 84-15898. Class No.: Y3.N88:10/1060.
OCLC: 10679044
 Report indicates that no workers who were present during
TMI-2 nuclear power plant emergencies were exposed to radiation
in amounts sufficient to be hazardous to their health. This NRC
report has particular significance for health professionals and local
environmental planners.

195. TMI-1 RESTART
 Clarification of TMI Action Plan Requirements. Requirements
for Emergency Response Capability. Supplement Number 1. Prepared by U.S. Nuclear Regulatory Commission. Washington, D.C.:
The Commission, 1983. 34p.
Mo. Cat.: 83-18283. Class No.: Y3.N88:10/0737/suppl.1.
OCLC: 9593036
 Document, from NRC director of the Division of Licensing,
lists new technical safety requirements for all U.S. commercial nuclear reactors now in operation. New requirements have been made
in light of TMI-2 accident findings.

196. TMI-2 FINANCIAL EFFECTS
 Authorization of Appropriations for the U.S. Nuclear Regulatory Commission for Fiscal Year 1982 and Fiscal Year 1983. Part 3.
Washington, D.C.: U.S. Govt. Print. Off., 1983. 315p. Bibl.
Mo. Cat.: 83-16605. Class No.: Y4.In8/14:97-25/pt.3.
CIS: 83-H441-23. OCLC: 8433909
 Budget hearing for FY 83 includes information on revised costs
of clean-up and decontamination activities at TMI-2 (pp. 60-92).
Technical information will have particular interest for scientists and
nuclear engineers.

197. TMI-2 ACCIDENT RECOVERY ACTIVITIES
 Current Status of Three Mile Island Nuclear Generating Station, Units 1 & s. Washington, D.C.: U.S. Govt. Print. Off.,
1983. 214p.
Mo. Cat.: 83-21484. Class No.: Y4.In8/14:98-10.
CIS: 83-H441-27. OCLC: 9771506
 Oversight hearing reviews progress of TMI-2 clean-up activities (pp. 3-28, 36-72, 104-191) and also describes TMI-1 repair
and proposed restart activities (pp. 28-35).

198. TMI-2 ACCIDENT RECOVERY ACTIVITIES
 Department of Energy Research and Development Programs.
Washington, D.C.: U.S. Govt. Print. Off., 1983. 895p. Bibl.
Mo. Cat.: 83-19943. Class No.: Y4.En2:97-112.
CIS: 83-S311-18. OCLC: 9617530
 DOE budget hearings for FY 83 contain an evaluation of progress of TMI-2 clean-up, with attention to nuclear reactor safety and
management of nuclear waste materials from the TMI-2 emergency
(pp. 215-280).

U.S. Government Hearings, etc. 93

199. POTASSIUM IODIDE, USES OF
Emergency Preparedness for Radiological Accidents: The Issue of Potassium Iodide. Washington, D.C.: U.S. Govt. Print. Off., 1983 [i.e. 1984]. 396p. Bibl.
Mo. Cat.: 84-13923. Class No.: Y4.In8/14:97-40.
CIS: 84-H441-5. OCLC: 10583556
Hearing examines feasibility of using potassium iodide (KI) as a thyroid-blocking agent for persons exposed to radiation during a dangerous nuclear power plant accident. The decision not to use KI during the TMI-2 emergency is explained (pp. 271-298).

200. TMI-2 ACCIDENT RECOVERY ACTIVITIES
Energy and Water Development Appropriations, Fiscal Year 1982. Part 2: Department of Energy. Washington, D.C.: U.S. Govt. Print. Off., 1983. numbered iii + 645-1287 p. + vii. Bibl.
Mo. Cat.: 83-23149. Class No.: Y4.Ap6/2:En2/2/983/pt.2.
CIS: 83-S181-4. OCLC: 10533832
DOE budget request for 1983 includes evaluations of TMI-2 accident clean-up and proposed research programs seeking to improve U.S. commercial nuclear power plant safety (pp. 1133-1287).

201. NUCLEAR POWER PLANT SAFETY
Energy Department Budget for Fiscal Year 1984. Washington, D.C.: U.S. Govt. Print. Off., 1983. 330p.
Mo. Cat.: 83-23194. Class No.: Y4.En2/3:98-20.
CIS: 83-H361-62. OCLC: 9859954
FY 84 budget hearings focus on proposed DOE nuclear programs, with detailed information on proposed nuclear safety research dealing with problems related to findings of TMI-2 accident.

202. TMI-2 RADIATION EFFECTS
EPA Radon and Radionuclide Emission Standards. Washington, D.C.: U.S. Govt. Print. Off., 1983. 657p. Bibl.
Mo. Cat.: 84-11650. Class No.: Y4.Ar5/2a:983-84/11.
CIS: 84-H201-10. OCLC: 10383201
Hearing receives supplementary information on EPA study on effects of radiation exposure, including a comparison of radiation exposure experienced during TMI-2 accident with radiation exposure experienced during routine medical tests (pp. 622-639).

203. TMI-2 ACCIDENT RECOVERY ACTIVITIES
Energy & Water Development Appropriations, Fiscal Year 1984. Part 2: Department of Energy. Washington, D.C.: U.S. Govt. Print. Off., 1983. numbered iii, 585-1312 p. + vii. Bibl.
Mo. Cat.: 84-7651. Class No.: Y4.Ap6/2:S.hrg.98-126/pt.2.
CIS: 83-S181-39.3. OCLC: 10755046
DOE budget request for FY 84 contains information on TMI-2 accident causes and progress of clean-up activities (pp. 814-936).

204. TMI-2 ACCIDENT RECOVERY ACTIVITIES
Evaluation of the Three Mile Island Unit 2 Reactor Building

Decontamination Process. Prepared by D. Dougherty and J.W. Adams, for U.S. Nuclear Regulatory Commission. Washington, D.C.: The Commission, 1983. 70p. Bibl.
Mo. Cat.: 84-11589. Class No.: Y3.N88:25/3381.
OCLC: 10439139
 Progress of clean-up at TMI-2 reactor building is assessed. Outlined are tasks designed to remove and to dispose safely of the remaining radioactive wastes from TMI-2.

205. REACTOR GAUGE, IMPROVEMENTS FOR
 Feasibility Study on the Development of a Non-invasive Liquid Level Gauge for Nuclear Power Reactors. Prepared by A.J. Baratta [et al.] for U.S. Nuclear Regulatory Commission. Washington, D.C.: The Commission, 1983. 76p. Bibl.
Mo. Cat.: 83-24764. Class No.: Y3.N88:25/3290.
OCLC: 9958162
 Technical report describes development of an improved pressure gauge for U.S. commercial light water reactors which has an improved ability to show density and level changes and decreased pressure. Such a gauge could have lessened the severity of the TMI-2 accident.

206. TMI-2 FINANCIAL EFFECTS
 Financing the Cleanup of Three Mile Island Unit 2 Nuclear Powerplant. Washington, D.C.: U.S. Govt. Print. Off., 1983. 610p. Bibl.
Mo. Cat.: 83-14604. Class No.: Y4.In8/14:97-36.
CIS: 83-H441-15. OCLC: 9368864
 Hearings consider various pending bills designed to assist with remaining clean-up costs related to TMI-2 accident and to provide adequate insurance protection in the event of future serious commercial nuclear power plant accidents.

207. TMI-2 ACCIDENT RECOVERY ACTIVITIES
 Fiscal Year 1984 Department of Energy Authorization (Nuclear Fission, R & D and Waste Management). Vol. 4. Washington, D.C.: U.S. Govt. Print. Off., 1983. 1044p. Bibl.
Mo. Cat.: 84-7762. Class No.: Y4.Sci2:98/18.
CIS: 83-H701-71. OCLC: 10724351
 Hearing receives information on TMI-2 clean-up, research and development activities, including roles of federal government, the electric utility industry, the GPU Corporation and its subsidiaries in the clean-up activities (pp. 876-934, 936-980).

208. NUCLEAR POWER PLANT REGULATION
 Nuclear Licensing Reform. Washington, D.C.: U.S. Govt. Print. Off., 1983. 370p. Bibl.
Mo. Cat.: 84-20143. Class No.: Y4.In8/14:98-25.
CIS: 84-H441-17. OCLC: 10859215
 Hearing includes information concerning various bills designed to revise licensing of U.S. nuclear power plants in the light of TMI-2 accident findings and subsequent regulations developed to make operation at commercial nuclear power plants safer.

U.S. Government Hearings, etc. 95

209. TMI-2 ACCIDENT RECOVERY ACTIVITIES
Nuclear Regulatory Commission's Budget Request for Fiscal Years 1984 and 1985. Washington, D.C.: U.S. Govt. Print. Off., 1983. 486p. Bibl.
Mo. Cat.: 83-21504. Class No.: Y4.P96/10:S.hrg.98-53.
CIS: 84-S321-15. OCLC: 9702012
Hearing receives NRC report on nuclear power plant licensing changes since the TMI-2 accident and progress of TMI-2 clean-up and decontamination activities (pp. 86-101).

210. NUCLEAR POWER PLANT SAFETY
Planning Guidance for Nuclear Power Plant Decontamination. Prepared by L.F. Munson [et al.] for U.S. Nuclear Regulatory Commission. Washington, D.C.: Division of Engineering, Office of Nuclear Reactor Regulation, U.S. Nuclear Regulatory Commission. 1983. 105p. Bibl.
Mo. Cat.: 84-15923. Class No.: Y3.N88:25/2963.
OCLC: 10509206
This NRC research report discusses general procedures to be used in planning for any future extensive radioactive decontamination activities at a U.S. commercial nuclear power plant experiencing a serious accident similar to the TMI-2 emergency. Special attention is given to procedures used for the TMI-2 emergency.

211. TMI-2 ENVIRONMENTAL EFFECTS
Programmatic Environment Impact Statement Related to Decontamination and Disposal of Radioactive Wastes Resulting from March 28, 1979 Accident, TMI Nuclear Station, Unit 2. Docket no. 50-320: Draft Supplement Dealing with Occupational Radiation Dose. Washington, D.C.: U.S. Nuclear Regulatory Commission, 1983. 1 vol. passim. Bibl.
Mo. Cat.: 84-17994. Class No.: Y3.N88:10/0683/draft/suppl.1.
OCLC: 10828542
Report describes environmental effects related to removal of radioactive waste materials from TMI-2 between July 1983 and October 1983. Technical report has interest for environmental planners and nuclear engineers. Report was developed by GPU Nuclear, Inc. for the use of the NRC.

212. TMI-1 RESTART
Safety Evaluation Report Related to Steam Generator Tube Repair and Return to Operation, TMI Nuclear Station, Unit no. 1. Docket 50-289, GPU Nuclear Corporation, et al. Prepared by Office of Nuclear Reactor Regulation. Washington, D.C.: U.S. Nuclear Regulatory Commission, 1983. 1 vol. passim. Bibl.
Mo. Cat.: 84-11553. Class No.: Y3.N88:10/1019.
OCLC: 10474646
GPU Corporation's efforts to bring TMI-1's steam generator to readiness for a safe restart are assessed. Particular attention is given to the proposed TMI-1 operational changes and new safety measures already in place. Technical report is developed for use of the NRC.

213. TMI-1 RESTART
TMl-1 Restart: An Evaluation of the RHR, BETA, and Draft INPO Reports as They Affect Restart Issues at TMI Nuclear Station, Unit 1. Docket 50-289. Washington, D.C.: U.S. Nuclear Regulatory Commission, 1983. 1 vol. passim. Bibl.
Mo. Cat.: 84-15856. Class No.: Y3.N88:10/0680/suppl.4.
OCLC: 10633524
Report summarizes various reports and findings and includes various recommendations pertaining to restart of TMI-1. Report is developed for use of the NRC.

214. NUCLEAR POWER PLANT SAFETY
Torsional Ultrasonic Technique for LWR Liquid Level Measurement. Prepared by W.B. Dress for U.S. Nuclear Regulatory Commission. Washington, D.C.: The Commission, 1983. 16p. Bibl.
Mo. Cat.: 83-24753. Class No.: Y3.N88:25/3113.
OCLC: 9877862
Technical report suggests method whereby an ultrasonic wave guide can be utilized as a sensor to improve the accuracy of nuclear reactor instrument reading and thereby decrease the chance that another serious U.S. commercial nuclear power plant accident, like the TMI-2 emergency, could occur.

215. TMI-2 ACCIDENT RECOVERY ACTIVITIES
Answers to Frequently Asked Questions About Cleanup Activities at Three Mile Island Unit 2. Revision 1. Prepared by TMI Program Office, U.S. Nuclear Regulatory Commission. Washington, D.C.: The Program Office, 1984. 56p.
Mo. Cat.: 84-17996. Class No.: Y3.N88:10/0732/rev.1.
OCLC: 10828315
Recent TMI-2 clean-up activities and related problems are presented, together with an assessment of the remaining clean-up tasks. Question and answer format makes this publication useful for a broad audience--i.e., students, beginning researchers, etc.

216. TMI-2 ACCIDENT
Energy and Water Development Appropriations for 1985. Part 5: U.S. Department of Energy. Washington, D.C.: U.S. Govt. Print. Off., 1984. 1113p.
Mo. Cat.: 84-16116. Class No.: Y4.Ap6/1:En2/2/985/pt.5.
CIS: 84-H181-26. OCLC: 10485612
Hearings for DOE FY 85 budget requests contain information on various nuclear safety research programs and include information on TMI-2 clean-up activities (pp. 429-641).

217. NUCLEAR POWER PLANT SAFETY
Comparison of Implementation of Selected TMI Action Plan Requirements at Operating Plants Designed by Babcock and Wilcox. Prepared by John O. Thoma. Washington, D.C.: U.S. Nuclear Regulatory Commission, 1984. 194p.
Mo. Cat.: 84-19968. Class No.: Y3.N88:10/1066.
OCLC: 10918812

U.S. Government Hearings, etc.   97

NRC technical report compares and contrasts safety procedures at U.S. commercial power facilities which have Babcock and Wilcox-designed reactors similar to the damaged TMI-2 reactor, following NRC publication of TMI-2 safety-related requirements for all U.S. commercial power facilities.

218. NUCLEAR POWER PLANT SAFETY
Energy and Water Development Appropriations for Fiscal Year 1985. Part 2. Washington, D.C.: U.S. Govt. Print. Off., 1984. 882p.
Mo. Cat.: 85-11676. Class No.: Y4.Ap6/2:S.hrg.98-911/pt.2.
CIS: 85-S181-4. OCLC: 11223198
Congressional hearing includes an evaluation of the impact of the TMI-2 accident findings upon nuclear safety-related research at U.S. commercial nuclear power facilities (pp. 729-747).

219. TMI-2 ACCIDENT RECOVERY ACTIVITIES
Energy and Water Development Appropriations for Fiscal Year 1985. Part 6. Washington, D.C.: U.S. Govt. Print. Off., 1984. 1556p.
Mo. Cat.: 84-16116. Class No.: Y4.Ap6/1:En2/2/985/pt.6.
CIS: 84-H181-27. OCLC: 10485612
NRC budget request for Fiscal Year 1985 contains general information regarding the NRC's safety recommendations for all U.S. commercial nuclear power facilities. Detailed information is presented on progress of TMI-2 decontamination and recovery activities (pp. 1283-1542).·

220. TMI-2 ACCIDENT RECOVERY ACTIVITIES
Evaluation of Nuclear Facility Decommissioning Projects: Summary Report, TMI Unit 2 Polar Crane Recovery. Prepared by D.H. Doerge and R.L. Miller for U.S. Nuclear Regulatory Commission. Washington, D.C.: Division of Engineering Technology, Office of Nuclear Regulatory Research, U.S. Nuclear Regulatory Commission, 1984. 61p.
Mo. Cat.: 85-11622. Class No.: Y3.N88:25/3884.
OCLC: 11571210
NRC technical report describes development of the polar crane, designed to assist with removal and subsequent examinations of damaged TMI-2 reactor core. Report provides considerable information concerning TMI-2 recovery operations, of use to engineers and researchers.

221. TMI-2 ACCIDENT RECOVERY ACTIVITIES
David Paul Hodel Nomination. Washington, D.C.: U.S. Govt. Print. Off., 1985. 528p. Bibl.
Mo. Cat.: 85-16225. Class No.: Y4.En2:S.hrg.99-6.
CIS: 85-S311-31. OCLC: 11895737
Congressional hearing, discussing proposed nomination of David P. Hodel as U.S. Secretary of the Interior, includes information on U.S. Dept. of Energy contract for containers designed for

98   Three Mile Island

storage and transport of radioactive waste materials from the damaged
TMI-2 reactor (pp. 477-483).

222. TMI-1 RESTART
      Fiscal Year 1985, Budget Review. Washington, D.C.: U.S.
Govt. Print. Off., 1984. 864p.
Mo. Cat.: 84-21705. Class No.: Y4.P96/10S.hrg.98-758.
CIS: 84-S321-27. OCLC: 11006107
      Congressional hearings include information on planning for
proposed TMI-1 restart by the company operators (pp. 180-198)
and differing testimony concerning the safety of the proposed TMI-1
restart (pp. 193-384).

223. TMI-1 RESTART
      Investigation of the Sulfur and Lithium to Sulfur Ratio Threshold in Stress Corrosion Cracking of Sensitized Alloy 600 in Borated Thiosulfate Solution. Prepared by R. Bandy and K. Kelly for U.S.
Nuclear Regulatory Commission. Washington, D.C.: Division of
Engineering Technology, Office of Nuclear Regulatory Research, U.S.
Nuclear Regulatory Commission, 1984. 36p. Bibl.
Mo. Cat.: 84-23033. Class No.: Y3.N88:25/3834.
OCLC: 11120393
      Technical report evaluates the effects of stress corrosion on
the steam boiler within the TMI-1 reactor during its enforced shutdown following the TMI-2 emergency. Report has relevance for
nuclear engineers and scientists concerned with the operation of
U.S. commercial power facilities.

224. NUCLEAR POWER PLANT SAFETY
      Management Weaknesses Affect Nuclear Regulatory Commission
Efforts to Address Safety Issues Common to Nuclear Power Plants.
Report to Congress. By the Comptroller General of the United States.
Washington, D.C.: U.S. General Accounting Office, [1984]. 9p.
Mo. Cat.: 85-8517. Class No.: GA1.13:RCED-84-149/sum.
OCLC: 11478768
      U.S. General Accounting Office Report indicates NRC administrative problems have slowed the NRC's efforts to enforce more
rigorous safety operational changes and standards at U.S. commercial nuclear power facilities following publication of TMI-2 accident
findings. Report is directed to members of Congress.

225. TMI-2 ACCIDENT RECOVERY ACTIVITIES
      Nuclear Regulatory Commission Budget Request for Fiscal Years
1984 and 1985. Washington, D.C.: U.S. Govt. Print. Off., 1984.
814p. Bibl.
Mo. Cat.: 85-5698. Class No.: Y4.In8/14:98-31.
CIS: 84-H441-4. OCLC: 11427641
      Congressional hearings contain information on progress of
TMI-2 clean-up operations (pp. 437-573) as well as information concerning alleged safety violations by TMI operators prior to the TMI-2 accident (pp. 706-720).

U.S. Government Hearings, etc. 99

226. TMI-2 ACCIDENT RECOVERY ACTIVITIES
NRC Plan for Cleanup Operations at Three Mile Island Unit 2.
Rev. 2. Prepared by TMI Program Office. Washington, D.C.:
U.S. Nuclear Regulatory Commission, 1984. 1 vol. passim. Bibl.
Mo. Cat.: 84-17995. Class No.: Y3.N88:10/0698/rev.2.
OCLC: 10788515
Technical report outlines revised NRC plan for removal of remaining radioactive waste materials from TMI-2 and related decontamination activities.

227. TMI-2 ACCIDENT RECOVERY ACTIVITIES
Review of a Test Program for Qualifying the Solidification of EPICOR-II Resins with Cement. Prepared by R.E. Barletta and R.E. Davis for U.S. Nuclear Regulatory Commission. Washington, D.C.: The Commission, 1984. 33p. Bibl.
Mo. Cat.: 84-16034. Class No.: Y3.N88:25/3496.
OCLC: 10692208
As a part of TMI-2 clean-up activities, a program for solidifying EPICOR-II liners in cement was tried in a laboratory setting. This technical report indicates progress but recommends further tests, as solidification of the liners was not accomplished satisfactorily.

228. TMI-2 ACCIDENT RECOVERY ACTIVITIES
Authorizing Appropriations to the Department of Energy for Civilian Energy Programs for Fiscal Year 1986 and Fiscal Year 1987, and for other Purposes: Report (to accompany 2041). (Incl. the cost estimate of the Congressional Budget Office). Washington, D.C.: U.S. Govt. Print. Off., 1985. 18p.
Mo. Cat.: 85-19977. Class No.: Y1.1/8:99-118/pt.1.
CIS: 85-H443-15. OCLC: 12205683
Congressional report includes proposed appropriations for various U.S. Department of Energy programs, including research projects to assist with TMI-2 recovery and decontamination activities (p. 2).

229. TMI-2 ACCIDENT RECOVERY ACTIVITIES
Programmatic Environmental Impact Statement Related to the Decontamination and Disposal of Radioactive Wastes Resulting from March 28, 1979 Accident TMI Nuclear Station, Unit 2. Docket no. 50-320: Final Supplement Dealing with Occupational Radiation Dose. GPU Nuclear, Inc. Washington, D.C.: U.S. Nuclear Regulatory Commission, 1984. 177p. Bibl.
Mo. Cat.: 84-17994. Class No.: Y3.N88:10/0683/suppl.1/Final rpt.
OCLC: 11741317
Supplementary report provides information indicating that personnel involved in TMI-2 clean-up and decontamination activities may be exposed to higher radiation doses than originally anticipated.

230. TMI-1 RESTART
TMI-1 Restart: An Evaluation of the Licensee's Management Integrity As It Affects Restart of TMI Nuclear Station, Unit 1. Docket

50-289. Washington, D.C.: U.S. Nuclear Regulatory Commission, 1984. 1 vol. passim. Bibl.
Mo. Cat.: 84-24481. Class No.: Y3.N88:10/0680/suppl.5.
OCLC: 11227248
Technical report favorably evaluates GPU Nuclear's petition to restart TMI-1, subject to the completion of various items enumerated in the report. This report has immediate use for GPU Nuclear, its parent company and the associated companies.

231. NUCLEAR POWER PLANT SAFETY
Better Inspection Management Would Improve Oversight of Operating Nuclear Power Plants: Summary Report to the Congress. Prepared by the Comptroller General of the United States. Washington, D.C.: U.S. General Accounting Office, [1985]. 9p.
Mo. Cat.: 85-19151. Class No.: GA1.13:RCED-85-5/sum.
OCLC: 12242041
Technical report suggests NRC inspection program for U.S. commercial facilities has improved since the TMI-2 emergency. However, the report suggests safety precautions are still inadequate and recommends additional training for plant inspectors and stricter safety regulations at U.S. commercial nuclear power facilities.

232. TMI-2 EMERGENCY PLANNING
Emergency Preparedness and the Licensing Process for Commercial Nuclear Power Reactors. Part 2. Washington, D.C.: U.S. Govt. Print. Off., 1985. 403p.
Mo. Cat.: 85-22315. Class No.: Y4.In8/14:98-52/pt.2.
CIS: 84-H441-35. OCLC: 12086506
Congressional hearing includes information on emergency planning of actual evacuation procedures carried out during the TMI-2 emergency, based on personal interviews and written surveys of the evacuees (pp. 271-76).

233. TMI-2 ACCIDENT RECOVERY ACTIVITIES
EPICOR-11 Resin Degradation Results from First Resin Samples of PF-8 and PF-20. Prepared by J. W. McConnell, Jr. and R.D. Sanders, Sr. Washington, D.C.: U.S. Nuclear Regulatory Commission, 1985. 48p. Bibl.
Mo. Cat.: 85-26541. Class No.: Y3.N88:25/4150.
OCLC: 12597800
NRC technical report describes proposed use of EPICOR-11 ion exchange resins in decontamination activities at TMI-2. Report deals with problems of radioactive waste management at TMI-2 and has particular interest for nuclear engineers, environmental planners, and researchers.

234. TMI-1 RESTART
Fiscal Year 1986 Budget Review. Washington, D.C.: U.S. Govt. Print. Off., 1985. 1049p. Bibl.
Mo. Cat.: 85-24018. Class No.: Y4.P96/10:S.hrg.99-69.
CIS: 85-S321-18. OCLC: 12337306

U.S. Government Hearings, etc. 101

Congressional hearings include information on proposed restart of TMI-1 and an assessment of these proposals by the NRC commissioners (pp. 253-269).

235.   TMI-2 ACCIDENT RECOVERY ACTIVITIES
Fiscal Year 1986 Department of Energy Authorization (Nuclear Fission R & D and Waste Management). Volume 4. Washington, D.C.: U.S. Govt. Print. Off., 1985. 803p.
Mo. Cat.: 86-2062. Class No.: Y4.Sci2:99-15.
CIS: 85-H701-93. OCLC: 12699048
Congressional hearings include summary of TMI-2 accident - recovery and decontamination activities in progress (pp. 11-15, 522-545). Remaining tasks which must be accomplished to complete the final TMI-2 clean-up process are noted.

236.   NUCLEAR POWER PLANT SAFETY
Lessons Learned in Utility Management and Status of the R & D Program Following the Accident at TMI. Washington, D.C.: U.S. Govt. Print. Off., 1985. 512p. Bibl.
Mo. Cat.: 85-13922. Class No.: Y4.Sci2:98/125.
CIS: 84-H701-41. OCLC: 11654742
Congressional hearing reviews U.S. commercial nuclear power plant safety changes made as a consequense of TMI-2 accident findings (pp. 250-277, 279-331). TMI-2 clean-up progress and proposed plans for restart of TMI-1 are noted.

237.   TMI-1 RESTART
NRC Management Policies. Washington, D.C.: U.S. Govt. Print. Off., 1985. 134p.
Mo. Cat.: 85-13825. Class No.: Y4.En2/3:98-166.
CIS: 85-H361-45. OCLC: 11777837
Congressional hearing contains information on proposed restart of TMI-1, with comments on the NRC's procedures for licensing U.S. commercial nuclear facilities (pp. 58-63).

238.   NUCLEAR POWER PLANT SAFETY
The Nuclear Regulatory Commission Should Report on the Progress in Implementing Lessons Learned from the Three Mile Island Accident Summary: Report to Congress. Prepared by the Comptroller General of the United States. Washington, D.C.: U.S. General Accounting Office, [1985]. 9p.
Mo. Cat.: 85-24928. Class No.: GA1.13:RCED-85-72/sum.
OCLC: 12590273
GAO report, issued July 19, 1985, indicates need for the NRC to present a detailed progress report regarding safety standards for U.S. commercial nuclear facilities developed as a consequence of TMI-2 accident findings.

239.   NUCLEAR POWER PLANT SAFETY
Nuclear Regulatory Reform. Washington, D.C.: U.S. Govt. Print. Off., 1985. 517p. Bibl.

102   Three Mile Island

Mo. Cat.: 86-5745. Class No.: Y4.P96/10:S.hrg.99-209.
CIS: 86-S321-1. OCLC: 12783697
    Congressional hearings include testimony concerning new safety
procedures and extended training in emergency procedures for plant
reactor operators made as a consequence of TMI-2 accident findings
(pp. 24-39, 126-190).

PART 3

U.S. GOVERNMENT PUBLICATIONS:
JOURNAL ARTICLES*

240. TMI-2, PRESIDENTIAL COMMISSION, APPOINTMENT OF
"Carter Orders Decontrol of Oil Prices, Three Mile Island
Probe." Energy Insider 2 (April 16, 1979): 4.
Class No.: E1.54:2-8. Item No.: 429-T-37.
Series/Report No.: Current U.S. Govt. Periodicals on Micro.,**
Order No. 79-245.
OCLC: 4280387
President Carter's appointment of a special presidential commission to investigate the causes of the TMI-2 accident is noted. Brief article has interest for public.

241. TMI-2 RADIATION MONITORING
"DOE Equipment, Radiological Teams at Three Mile Island Site."
Energy Insider 2 (April 16, 1979): 1.
Class No.: E1.54:2-8. Item No.: 429-T-37.
Series/Report No.: Order No. 79-245. OCLC: 4280387
Announcement is made of various U.S. Department of Energy activities to monitor possible radiation releases into the air or water as a result of the TMI-2 accident. Article has general interest for public.

242. TMI-2 RADIATION MONITORING
"NWS Helped During Nuclear Plant Accident." NOAA 9 (July 1979): 61.
Class No.: C55.14:9-3. Item No.: 250-E-1.
Series/Report No.: Order No. 79-143. OCLC: 1768577
U.S. National Weather Service activities during the TMI-2 emergency are described, emphasizing air monitoring for possible radiation releases. Brief report has general public interest.

243. TMI-2 RADIATION MONITORING
"Three Mile Island Data Set." EDIS: Environmental Data and Information Service 10 (July 1979): 24.

---

*First line of each entry consists of entry number and subject of the entry. Title and bibliographic details follow on subsequent lines.
**Note: Full title is U.S. Government Periodicals on Microfiche; hereafter title is omitted, and only the order number is given.

Class No.: C55.222:10-4. Item No.: 273-D-7.
Series/Report No.: Order No. 79-215. OCLC: 4359560
Brief report, with interest for the general public, suggests no long-term environmental problems are expected as a result of the TMI-2 accident. Conclusion is based on monitoring of air and water samples taken during and following the TMI-2 emergency.

244. TMI-2, PUBLIC REACTION TO ACCIDENT
"Mayday at Three Mile Island," Postal Life 8 (July/August 1979): 11.
Class No.: P1.43:8-4. Item No.: 838-D.
Series/Report No.: Order No. 79-088. OCLC: 1762710
Postal workers living near TMI power plant describe their immediate reactions to the TMI-2 crisis and the potential dangers of radiation releases from the accident. Article has general public interest.

245. TMI-2 HEALTH STUDIES
"Population Dose and Impact on Health of Three Mile Island Accident." Public Health Reports 94 (July/August 1979): 387.
Class No.: HE20.6011:94-4. Item No.: 497.
Series/Report No.: Order No. 79-146. OCLC: 1799423
Brief article summarizes findings of report compiled by staff scientists from HEW, EPA and the NRC indicating four estimates of radiation released by the TMI-2 accident and possible health effects. Article has general public interest.

246. TMI-2 ACCIDENT REPORTS
"Preliminary Report on the Three Mile Island Incident," William R. Castro and William B. Cottrell. Nuclear Safety 20 (July/August 1979): 483-490. Bibl.
Class No.: Y3.N88.9/20-4. Item No.: 1051-H.
Series/Report No.: Order No. 79-079. OCLC: 1760906
Causes and chronological development of the TMI-2 accident are presented. A summary of NRC recommendations is given for the U.S. nuclear power facilities based on preliminary TMI-2 accident studies. Report is directed to scientists and engineers.

247. TMI-2 RADIATION MONITORING
"Primer on Radiation," Bill Rados. FDA Consumer 13 (July/August 1979): 4-9.
Class No.: HE20.4010:13-4. Item No.: 475-H.
Series/Report No.: Order No. 79-045. OCLC: 1685864
General public interest article describes radiation effects upon the human body, emphasizing the TMI-2 accident. As the human embryo and young children are considered most susceptible to radiation exposure, pregnant women and young children were advised to leave the TMI vicinity during the TMI-2 emergency.

248. NUCLEAR POWER PLANT SAFETY
"Resources, Technology and the Environment," Russell W. Peterson [et al]. EPA Journal 5 (September 1979): 6-8.

Class No.: EP1.67:5-8. Item No.: 431-1-66.
Series/Report No.: Order No. 79-200. OCLC: 2663078
  A member of the special TMI-2 presidential commission suggests there will be a decline in the number of U.S. nuclear power plants opened, as a result of public anxiety about TMI-2 accident dangers and the high cost of new plant construction incorporating TMI-2 related safety features. Interview has general public interest.

249. TMI-2 ACCIDENT REPORTS
  "Developments Pertaining to the Three Mile Island Accident," William B. Cottrell. Nuclear Safety 20 (September/October 1979): 613-623. Bibl.
Class No.: Y3.N88:9/20-5. Item No.: 1051-H.
Series/Report No.: Order No. 79-079. OCLC: 1760906
  Detailed summary of immediate causes and development of the TMI-2 accident is presented. Article has interest for engineers, scientists and the general public.

250. TMI-2 HEALTH STUDIES
  "Preliminary Dose and Health Impact of Accident at Three Mile Island Nuclear Station." Nuclear Safety 20 (September/October 1979): 591-594.
Class No.: Y3.N88:9/20-5. Item No.: 1051-H.
Series/Report No.: Order No. 79-079. OCLC: 1760906
  Health studies indicate population living in vicinity of Three Mile Island experienced no long-term adverse health effects as a result of the radioactive releases during the TMI-2 accident. Article is directed to public and to scientists.

251. NUCLEAR POWER PLANT REGULATION
  "Three Mile Island--Possible Regulatory Responses and Impact on Nation's Short-term Electric Utility Fuel Outlook," Gene Clark et al. Monthly Energy Review (Oct. 1979): i-viii.
Class No.: E3.9:79-10. Item No.: 434-A-2.
Series/Report No.: Order No. 79-163. OCLC: 1798576
  Various regulatory options and their implications for the U.S. nuclear power industry are considered, shortly after the TMI-2 accident. Report has particular interest for engineers and scientists.

252. NUCLEAR POWER PLANT SAFETY
  "Commissioners Speak." Nuclear Safety 20 (November/December 1979): 755-756.
Class No.: Y3.N88:9/20-6. Item No.: 1051-H.
Series/Report No.: Order No. 79-079. OCLC: 1760906
  Summary of four speeches pertaining to nuclear power and nuclear safety by NRC commissioners. All speeches make reference to the TMI-2 accident and need for greater U.S. nuclear power plant safety.

253. INSTITUTE OF NUCLEAR POWER OPERATIONS (INPO)
  "Institute of Nuclear Power Operations." Nuclear Safety 20 (November/December 1979): 756-757.

Class No.: Y3.N88:9/20-6. Item No.: 1051-H.
Series/Report No.: Order No. 79-079. OCLC: 1760906
General interest article describes the proposed INPO, designed to monitor and to assist with the improved operations of U.S. nuclear power reactors. INPO is sponsored by the Electric Power Research Institute (EPRI) for the electric utility industry.

254. NUCLEAR POWER PLANT SAFETY
"ACRS on Pipe Cracking and Safety Improvement," William B. Cottrell. Nuclear Safety 20 (November/December 1979): 752-753.
Class No.: Y3.N88:9/20-6. Item No.: 1050-H.
Series/Report No.: Order No. 79-079. OCLC: 1760906
Following the TMI-2 accident, a list of recommended technical procedures to improve operational safety at U.S. nuclear power facilities is presented by NRC Advisory Committee on Nuclear Safeguards. Recommendations are of direct interest to scientists and engineers.

255. TMI-2 ACCIDENT REPORTS
"Summary of TMI-2 Lessons Learned Task Force Report." Adapted by Nuclear Safety Staff. Nuclear Safety 20 (November/December 1979): 735-741. Bibl.
Class No.: Y3.N88:9/20-6. Item No.: 1051-H.
Series/Report No.: Order No. 79-079. OCLC: 176096
Summary of technical report presents recommendations for improved safety operations of U.S. nuclear power plants in light of TMI-2 accident findings. Report has interest for the general public and scientists.

256. TMI-2 LEGAL PROBLEMS
"TMI-2 An Extraordinary Nuclear Occurrence?" Nuclear Safety 20 (November/December 1979): 753-755.
Class No.: Y3.N88:9/20-6. Item No.: 1051-H.
Series/Report No.: Order No. 79-079. OCLC: 1760906
Summary is presented of general-interest information regarding lawsuits and claims involving personal injury and property damage resulting from the TMI-2 accident. The NRC's discretionary power to determine the seriousness of the TMI-2 accident is discussed.

257. NUCLEAR POWER PLANT REGULATION
"Commissioners Speak." Nuclear Safety 21 (January/February 1980): 125-126.
Class No.: Y3.N88:9/21-1. Item No.: 1051-H.
Series/Report No.: Order No. 80-079. OCLC: 1760906
Summary of speech by NRC Commissioner Gilinsky indicates decline in orders for nuclear power reactors began before the TMI-2 accident. Increased regulations for U.S. operators of nuclear power facilities since TMI-2 accident are noted.

258. NUCLEAR POWER PLANT REGULATION
"Post-TMI Policy on Nuclear Plant Licensing," William B. Cottrell. Nuclear Safety 21 (January/February 1980): 124.

Journal Articles 107

Class No.: Y3.N88:9/21-1. Item No.: 1051-H.
Series/Report No.: Order No. 80-079. OCLC: 1760906
General interest report notes changes in NRC licensing procedures for U.S. commercial nuclear reactors following the TMI-2 accident.

259. EMERGENCY COMMUNICATIONS SYSTEMS
"Technical Assessment of Disturbance Analysis Systems," A.B. Long. Nuclear Safety 21 (January/February 1980): 38-50. Bibl.
Class No.: Y3.N88:9/21-1. Item No.: 1051-H.
Series/Report No.: Order No. 80-079. OCLC: 1760906
In view of the high level of noise produced by various alarm systems during the initial stages of the TMI-2 accident, this report suggests ways to improve warning systems for accidents. Report has special significance for nuclear engineers.

260. TMI-2 RADIATION RELEASES
"For Rationality in Radiation Risk Studies," Harold H. Rossi. Journal of the National Cancer Institute (JNCI) 64 (March 1980): 403.
Class No.: HE.20.3161:64-3. Item No.: 488.
Series/Report No.: Order No. 80-060. OCLC: 1064763
Letter, from Harold Rossi, Department of Radiology, Columbia University, states why the author believes radiation studies, for population living in vicinity of TMI during the TMI-2 accident, will produce inconclusive results. Letter has particular interest for scientists and health professionals.

261. NUCLEAR POWER PLANT REGULATION
"ACRS on the 'Licensing Pause,'" William B. Cottrell. Nuclear Safety 21 (March/April 1980): 261-262.
Class No.: Y3.N88:9/21-2. Item No.: 1051-H.
Series/Report No.: Order No. 80-079. OCLC: 1760906
The NRC Advisory Committee on Nuclear Safeguards outlines technical procedures designed to shorten the lengthy licensing process for nuclear plants without sacrificing health or safety considerations. Article is directed to engineers and scientists.

262. NUCLEAR POWER PLANT REGULATION
"Commissioners Speak." Nuclear Safety 21 (March/April 1980): 263-264.
Class No.: Y3.N88:9/21-2. Item No.: 1051-H.
Series/Report No.: Order No. 80-079. OCLC: 1760906
Summary of three speeches by NRC commissioners dealing with nuclear power and the lessons and problems raised by the TMI-2. All speeches have general public interest.

263. TMI-2 ACCIDENT REPORTS
"Report of the President's Commission on Accident At Three Mile Island." Adapted by Nuclear Safety Staff. Nuclear Safety 21 (March/April 1980): 234-244.

Class No.: Y3.N88:91/21-2. Item No.: 1051-H.
Series/Report No.: Order No. 80-079. OCLC: 1760906
 Article summarizes the important findings and recommendations of the President's Commission on Three Mile Island. Of interest to general public, scientists, engineers, etc.

264. TMI-2 ACCIDENT REPORTS
 "Rogovin Report on Three Mile Island 2." Nuclear Safety 21 (May/June 1980): 389-393.
Class No.: Y3.N88.9/21-3. Item No.: 1051-H.
Series/Report No.: Order No. 80-079. OCLC: 1760906
 Articles summarizes the conclusions of the special NRC inquiry designed to review the TMI-2 accident and, in particular, to assess the NRC's role during the accident. Both the general public and scientists will find useful information in this report.

265. NUCLEAR POWER PLANT REGULATION
 "Commissioners Speak." Nuclear Safety 21 (July/August 1980): 532-533.
Class No.: Y3.N88:9/21-4. Item No.: 1051-H.
Series/Report No.: Order No. 80-079. OCLC: 1760906
 Summary of Commissioner Hendrie's speech deals with the NRC's operational and political problems since the TMI-2 accident. Speech has general interest for public and scientists.

266. NUCLEAR POWER PLANT REGULATION
 "Seventy Nuclear Power Plants Now Operating in U.S., George W. Cunningham. Energy Insider 3 (August 4, 1980): 3-5.
Class No.: E1.54:3-8. Item No.: 429-T-37.
Series/Report No.: Order No. 80-425. OCLC: 4280387
 Overview of U.S. nuclear power facilities in 1980 suggests to DOE official that operational and safety problems related to the TMI-2 accident are being met and nuclear power will continue as an important energy source. Article has general interest for the public.

267. NUCLEAR POWER PLANT REGULATION
 "NRC Interim Policy on Nuclear Power Plant Accident Considerations." Nuclear Safety 21 (September/October 1980): 665.
Class No.: Y3.N88.9/21-5. Item No.: 1051-H.
Series/Report No.: Order No. 80-079. OCLC: 1760906
 General public interest article describes need for review of U.S. nuclear power plants' emergency preparedness procedures, based upon the preliminary studies of the TMI-2 accident.

268. TMI-2 RADIATION MONITORING
 "Spectrum of Missions." EPA Journal 6 (October 1980): 21-22, 24.
Class No.: EP1.67:6-9. Item No.: 431-I-66.
Series/Report No.: Order No. 80-200. OCLC: 2663078
 General interest report describes environmental monitoring

Journal Articles 109

activities of the EPA, with particular attention to the work of the Environmental Monitoring Systems Laboratory, Las Vegas, Nevada in testing air and water samples from the TMI-2 accident site in 1979.

269. NUCLEAR POWER PLANT SAFETY
"Safety Objectives for Nuclear Power Plants." Nuclear Safety 21 (November/December 1980): 690.
Class No.: Y3.N88:9/21-6. Item No.: 1051-H.
Series/Report No.: Order No. 80-079. OCLC: 1760906
Summary of speech by NRC Commissioner Joseph M. Hendrie describes new safety and operational procedures designed for U.S. commercial nuclear power facilities following the TMI-2 accident. Speech has general public interest.

270. NUCLEAR POWER PLANT SAFETY
"Technical Note: New Issues in Reactor Safety," Saul Levine. Nuclear Safety 21 (November/December 1980): 718-723. Bibl.
Class No.: Y3.N88.9/21-6. Item No.: 1051-H.
Series/Report No.: Order No. 80-079. OCLC: 1760906
Review of U.S. nuclear power plant safety, 1970-1980, with particular reference to the TMI-2 accident. Report has interest for the general public and the scientific community.

271. TMI-2 ACCIDENT REPORTS
"Three Mile Island Nuclear Accident," James C. Hatcher. GAO Review 15 (Winter 1980): 26-28.
Class No.: GA1.15:15-4. Item No.: 544-A *Rev.).
Series/Report No.: Order No. 80-054. OCLC: 1570360
Report describes development of TMI-2 accident and its aftermath, stressing General Accounting Office's role to audit the accuracy of various TMI-2 related governmental reports as these are completed. Article is directed to the general public.

272. NUCLEAR POWER PLANT SAFETY
"Nuclear Energy is Safe and Necessary," George Cunningham. Nuclear Safety 22 (January/February 1981): np.
Class No.: Y3.N88.9/22-1. Item No.: 1051-H.
Series/Report No.: Order No. 81-079. OCLC: 1760906
Brief general interest report, from the Ass't. Secretary for Nuclear Energy, suggests nuclear energy will supply a growing share of the U.S.'s electrical power. In spite of TMI-2 problems, he predicts research and improved operational standards will make U.S. nuclear power plants safer.

273. NUCLEAR SAFETY ANALYSIS CENTER
"Nuclear Safety Analysis Center," Edwin L. Zebroski and Miles C. Leverett. Nuclear Safety 22 (January/February 1981): 1-9. Bibl.
Class No.: Y3.N88.9/22-1. Item No.: 1051-H.
Series/Report No.: Order No. 81-079. OCLC: 1760906
Article, of interest to scientists and the general public, describes

the Nuclear Safety Analysis Center of the Institute of Nuclear Power Operations, emphasizing the Center's assistance with TMI-2 recovery activities and in developing new safety procedures for all U.S. nuclear power plants.

274. TMI-2, PUBLIC REACTION TO ACCIDENT
"Public Attitudes and Information on the Nuclear Option," Morris W. Firebaugh. Nuclear Safety 22 (March/April 1981): 147-156. Bibl.
Class No.:   Y3.N88:9/22-2.   Item No.:   1051-H.
Series/Report No.: Order No. 81-079. OCLC: 1760906
Summary is presented of findings of public opinion polls dealing with nuclear energy both before and after the TMI-2 accident. Article has interest for the general public, scientists and psychologists.

275. NUCLEAR POWER PLANT SAFETY
"Human Factors Engineering in the U.S. Nuclear Arena," Edward W. Hagen and G.T. Mays. Nuclear Safety 22 (May/June 1981): 337-347. Bibl.
Class No.:   Y3.N88:9/22-3.   Item No.:   1051-H.
Series/Report No.: Order No. 81-079. OCLC: 1760906
Technical report evaluates proposed operational changes designed to improve safety at U.S. nuclear power plants, emphasizing TMI-2 accident findings. Report is directed to engineers and scientists.

276. TMI-2 ACCIDENT RECOVERY ACTIVITIES
"Semiscale Program Summary: A Review of Mod-3 Results," T.K. Larson and E. A. Harvego. Nuclear Safety 22 (May/June 1981): 312-327. Bibl.
Class No.:   Y3.N88:9/22-3.   Item No.:   1051-H.
Series/Report No.: Order No. 81-079. OCLC: 1760906
Technical report, of particular interest to engineers and scientists, contains detailed description of semiscale experiments which have proved helpful in evaluating TMI-2 recovery procedures and in obtaining data for future computer code assessment.

277. TMI-2 ACCIDENT RECOVERY ACTIVITIES
"Accident Simulation with TRAC," John C. Vigil and Richard J. Pryor. Los Alamos Science 2 (Summer/Fall 1981): 36-53. Bibl.
Class No.:   E1.96:2-3.   Item No.:   429-A-12.
Series/Report No.: Order No. 81-328. OCLC: 6818444
Report describes simulation of TMI-2 accident development, using specially designed computer code TRAC to predict condition of damaged TMI-2 reactor. Engineers and scientists will find useful information in this report.

278. NUCLEAR POWER PLANT SAFETY
"Keeping Reactors Safe from Sabotage," William Bradley [et al]. Los Alamos Science 2 (Summer/Fall 1981): 120-131. Bibl.

Class No.: E1.96:2-2. Item No.: 429-A-12.
Series/Report No.: Order No. 81-328. OCLC: 6818444
Technical report focuses on safeguards against possible sabotage in a nuclear power facility, with special attention to TMI-2 accident findings. Report has particular interest for engineers and scientists.

279. NUCLEAR POWER PLANT SAFETY
"Primer on Reactor Safety Analysis," Michael O. Stevenson and James F. Jackson. Los Alamos Science 2 (Summer/Fall 1981): 5-25. Bibl.
Class No.: E1.96:2-3. Item No.: 429-A-12.
Series/Report No.: Order No. 81-328. OCLC: 6818444
Technical report, of particular interest to nuclear engineers and scientists, focuses on reactor safety in light of TMI-2 accident findings.

280. TMI-2 ACCIDENT RECOVERY ACTIVITIES
"Three Mile Island: Aftermath and Impact," Jay E. Boudreau. Los Alamos Science 2 (Summer/Fall 1981): 92-97.
Class No.: E1.96:2-3. Item No.: 429-A-12.
Series/Report No.: Order No. 81-328. OCLC: 6818444
Report gives overview of TMI-2 accident, the enormous cleanup problems, and the legal and environmental problems facing TMI-2's owners and other U.S. commercial nuclear power plants. Both the general public and scientists will find this report useful and informative.

281. NUCLEAR POWER PLANT SAFETY
"Three Mile Island and Multiple-failure Accidents," John R. Ireland [et al.]. Los Alamos Science 2 (Summer/Fall 1981): 74-91. Bibl.
Class No.: E1.96:2-3. Item No.: 429-A-12.
Series/Report No.: Order No. 81-328. OCLC: 6818444
Detailed technical report summarizes nuclear reactor failure problems, emphasizing TMI-2 problems during and following the accident. Report is designed for nuclear engineers and scientists.

282. NUCLEAR POWER PLANT SAFETY
"View from San Diego: Harold Agnew Speaks Out," Harold Agnew and Barb Bulkin. Los Alamos Science 2 (Summer/Fall 1981): 152-159.
Class No.: E1.96:2-3. Item No.: 429-A-12.
Series/Report No.: Order No. 81-328. OCLC: 6818444
Prominent U.S. scientist suggests TMI-2 accident has caused other U.S. commercial nuclear power plants to improve safety and operational standards. Agnew suggests gas-cooled may be safer then water-cooled reactors. Interview has particular interest for scientists and engineers.

283. NUCLEAR POWER PLANT SAFETY
"Defense-in-Depth Approach to Safety in Light of Three Mile

Island Accident," R.J. Breen. Nuclear Safety 22 (September/ October 1981): 561-569. Bibl.
Class No.: E1.93:22-5.* Item No.: 1051-H
Series/Report No.: Order No. 81-079. OCLC: 1760906
Technical article describes new nuclear power plant safety measures with special reference to the TMI-2 accident and to subsequent activities by the NRC and the electric power industry. Article has particular interest for engineers and scientists.

284. TMI-2 RADIATION MONITORING
"Radiation Risks and Morality," Margaret N. Maxey. Nuclear Safety 22 (September/October 1981): 560.
Class No.: E1.93:22-5. Item No.: 1051-H.
Series/Report No.: Order No. 81-079. OCLC: 1760906
Following the TMI-2 emergency and the radiation dangers, the author questions the ethical basis for locating nuclear power facilities in heavily populated areas. The article has general public interest.

285. TMI-2 ACCIDENT REPORTS
"Thermal-Hydraulic and Core-Damage Analysis of the TMI-2 Accident," John R. Ireland [et al.]. Nuclear Safety 22 (September/ October 1981): 583-593. Bibl.
Class No.: E1.93:22-5. Item No.: 1051-H.
Series/Report No.: Order No. 81-079. OCLC: 1760906
Technical report summarizes analyses of the TMI-2 accident, indicating maximum core damage and amount of hydrogen produced during first three hours of the accident. Engineers and scientists will find this article useful.

286. TMI-2 RADIATION MONITORING
"EPA's Role at Three Mile Island," Cristine Perham. EPA Journal 6 (October 1980): 18-20+.
Class No.: EP1.67:6-9. Item No.: 431-1-66.
Series/Report No.: Order No. 80-200. OCLC: 2663078
General public interest article describes air monitoring programs and water sampling testing undertaken by the EPA as a consequence of the TMI-2 emergency, together with preliminary test results.

287. TMI-2 HEALTH STUDIES
"Evolution and Current State of Radioiodine Control," J. Louis Kovach. Nuclear Safety 23 (January/February 1982): 44-56. Bibl.
Class No.: E1.93:23-1. Item No.: 1051-H.
Series/Report No.: Order No. 82-079. OCLC: 1760906
Technical report has information of special interest to engineers and scientists. Iodine control technology is evaluated with particu-

---

*Note: Issues Jan./Feb. 1975 to July/Aug. 1981 are classed Y3.N88.9/. Issues Sept./Oct. 1981 to present are classed E1.93:

lar reference to scientific observations made during and after the TMI-2 accident.

288. NUCLEAR POWER PLANT REGULATION
"Perspectives on the Regulation of Nuclear Power: A Time for Action," Nunzio J. Palladino. Nuclear Safety 23 (January/February 1982): 102-103.
Class No.: E1.93:23-1. Item No.: 1051-H.
Series/Report No.: Order No. 82-079. OCLC: 1760906
General interest speech by NRC chairman emphasizes the NRC's role in accelerating clean-up activities at TMI-2 and in improving operational safety at U.S. commercial nuclear power plants.

289. NUCLEAR POWER PLANT SAFETY
"Aerosol Filtration," Melvin W. First and Humphrey Gilbert. Nuclear Safety 23 (March/April 1982): 167-183.
Class No.: E1.93:23-2. Item No.: 1051-H
Series/Report No.: Order No. 82-074. OCLC: 1760906
Technical report describes aerosol filtration in U.S. nuclear power facilities 1968-1980, with special reference to the TMI-2 loss-of-coolant accident. Report has special significance for engineers and scientists.

290. TMI-1 RESTART
"Three Mile Island Restart," William B. Cottrell. Nuclear Safety 23 (March/April 1982): 221-222.
Class No.: E1.93:23-2. Item No.: 1051-H.
Series/Report No.: Order No. 82-079. OCLC: 1760906
Brief general interest report comments on planning connected with proposed resumption of operations at TMI-1. Mention is made of August 27, 1981 NRC decision to permit restart of TMI-1 reactor, provided listed corrections are completed.

291. TMI-2 ACCIDENT RECOVERY ACTIVITIES
"Three Mile Island Accident Recovery," William B. Cottrell. Nuclear Safety 23 (March/April 1982): 222-223.
Class No.: E1.93:23-2. Item No.: 1051-H.
Series/Report No.: Order No. 82-079. OCLC: 1760906
Article, of general public interest, notes progress of TMI-2 clean-up, including successful removal of first fully loaded zeolite liner from the TMI-2 containment building. Leaks in steam generator tubes at TMI-1 reactor do not change the author's opinion that TMI-1 can be safely operated.

292. NUCLEAR POWER PLANT SAFETY
"Technical Note: The Control Room Design Review," E.W. Hagen. Nuclear Safety 23 (May/June 1982): 291-299. Bibl.
Class No.: E1.93:23-3. Item No.: 1051-H.
Series/Report No.: Order No. 82-079. OCLC: 1760906
Report summarizes findings of studies pertaining to control room design to assist plant operators in controlling or preventing

future nuclear emergencies like the TMI-2 accident. Nuclear engineers will find this report particularly useful.

293. NUCLEAR POWER PLANT REGULATION
"Financial Paralysis of 'Over-Regulated' Utilities Is Responsible for Problems of Nuclear Power Industry," W. Kenneth Davis. Energy Insider 5 (June 1982): 2.
Class No.: E1.54:5-6. Item No.: 429-T-37.
Series/Report No.: Order No. 82-245. OCLC: 4280387
Article, of general public interest, describes financial problems of U.S. commercial nuclear power plants, to which must be added new safety and operational equipment designed to reduce dangers of another serious accident TMI-2.

294. TMI-2, PUBLIC REACTION TO ACCIDENT
"Post-TMI Perceived Risk from Nuclear Power in Three Communities," D.T. Manning. Nuclear Safety 23 (July/August 1982): 379-384. Bibl.
Class No.: E1.93:23-4. Item No.: 1051-H.
Series/Report No.: Order No. 82-079. OCLC: 1760906
Summary of telephone survey findings regarding public views of nuclear power safety. Responses appear to be conditioned by place of residence. General public and researchers will find this information useful.

295. TMI-1 RESTART
"Three Mile Island I Restart Experiences Delays," E.G. Silver. Nuclear Safety 23 (July/August 1982): 469.
Class No.: E1.93:23-4. Item No.: 1051-H.
Series/Report No.: Order No. 82-079. OCLC: 1760906
Report summarizes TMI-1's steam generator corrosion problems which have slowed proposed TMI-1 restart activities. Article has interest for the general public.

296. TMI-2 ACCIDENT RECOVERY ACTIVITIES
"Three Mile Island 2 Clean-up Continues," E.G. Silver. Nuclear Safety 23 (July/August 1982): 463-469.
Class No.: E1.93:23-4. Item No.: 1051-H.
Series/Report No.: Order No. 82-079. OCLC: 1760906
Clean-up activities at TMI-2 are noted, with emphasis on removal of radioactive water from the TMI-2 containment building. Article has general interest for public.

297. INSTITUTE OF NUCLEAR POWER OPERATIONS (INPO)
"Overview of INPO and Its Programs," E.K. Russ. Nuclear Safety 23 (November/December 1982): 635-642. Bibl.
Class No.: E1.93:23-6. Item No.: 1051-H.
Series/Report No.: Order No. 82-079. OCLC: 1760906
Article presents summary of nuclear safety programs initiated and carried on by Institute of Nuclear Power Operations (INPO) since the TMI-2 accident. Article has interest for general public and scientific community.

Journal Articles 115

298. NUCLEAR POWER PLANT SAFETY
"Potential for Reducing the Predicted Consequences of Nuclear Power Plant Accidents," Robert L. Civiak. Congressional Research Service Review 4 (May 1983): 16-18.
Class No.: LC14.19:4-5. Item No.: 807-A-5.
Series/Report No.: Order No. 83-314. OCLC: 4383881
In light of the TMI-2 accident findings, this article suggests that improved operational and safety standards at U.S. nuclear power plants has reduced both the probability and the dangers of another serious nuclear power plant accident. Article is directed to members of Congress and has general interest for the public.

299. TMI-2 LEGAL EFFECTS
"Nuclear Liability Insurance--the Price-Anderson Reparations System and Claims Experience of Nuclear Industry," Joseph Marrone. Nuclear Safety 24 (November/December 1983): 783-791. Bibl.
Class No.: E1.93:24-6. Item No.: 1051-H.
Series/Report No.: Order No. 83-079. OCLC: 1760906
Review of operations of Price-Anderson Law with special reference to TMI-2 accident. By 1983 total liability and loss paid as a result of the Price-Anderson reparation system for the TMI-2 accident claims totaled $29,028,445. Litigation is still in progress. Lawyers and government officials will find this information summary useful.

300. TMI-1 RESTART
"Three Mile Island I Restart: Slow Progress," E.G. Silver. Nuclear Safety 24 (November/December 1983): 859.
Class No.: E1.93:24-6. Item No.: 1051-H.
Series/Report No.: Order No. 83-079. OCLC: 1760906
Problems in restart of TMI-1 are summarized, with emphasis on NRC decision to permit testing of steam operators of TMI-1 and attention to various legal and administrative problems concerning the restart of TMI-1. Both government officials and the public will find this article useful.

301. TMI-2 HEALTH STUDIES
"Three Mile Island Population Registry," Marilyn Goldhaber [et al]. Public Health Reports 98 (November/December 1983): 603-609. Bibl.
Class No.: HE20.6011:98-6. Item No.: 497.
Series/Report No.: Order No. 83-146. OCLC: 1799423
Report summarizes development of TMI Census, the detailed collection of demographic data for persons living in vicinity of TMI-2 accident site, made in 1979. These data are expected to be basis for further long-term health related studies. Article has interest for scientists and the general public.

302. TMI-2 ACCIDENT RECOVERY ACTIVITIES
"TMI-2 In-vessel Sampling and Safety Review." Nuclear Safety 24 (November/December 1983): 860.
Class No.: E1.93:24-6. Item No.: 1051-H.
Series/Report No.: Order No. 83-079. OCLC: 1760906

Article, of interest to the general public, describes ongoing clean-up activities at TMI-2, focusing on the development of the polar crane to assist in examination of the damaged reactor core and removal of fuel debris.

303. NUCLEAR POWER PLANT SAFETY
"Hydrogen Behavior in Light-Water Reactors," Marshall Berman and John C. Cummings. Nuclear Safety 25 (January/February 1984): 53-74. Bibl.
Class No.: E1.93:25-1. Item No.: 1051-H.
Series/Report No.: Order No. 84-079. OCLC: 1760906
The generation of hydrogen during a light water reactor accident (LWR) is described, with special reference to the TMI-2 accident. Related research projects are noted. Report has most interest for engineers and scientists.

304. TMI-2 ACCIDENT RECOVERY ACTIVITIES
"Progress in the Recovery Operations at Three Mile Island Unit 2," George Kalman and Richard A. Weller. Nuclear Safety 25 (January/February 1984): 88-113. Bibl.
Class No.: E1.93:25-1. Item No.: 1051-H.
Series/Report No.: Order No. 84-079. OCLC: 1760906
An evaluation of various TMI-2 recovery activities, with emphasis on work connected with defueling the damaged reactor core. Article has interest for public and scientists.

305. TMI-2 ACCIDENT RECOVERY ACTIVITIES
"TMI-2 Recovery Work Continues," E.G. Silver. Nuclear Safety 25 (January/February 1984): 124-126.
Class No.: E1.93:25-1. Item No.: 1051-H.
Series/Report No.: Order No. 84-079. OCLC: 1760906
Report focuses on 1983 TMI-2 clean-up progress and related problems. Summary of NRC reports pertaining to TMI-2 and legal and administrative problems of GPU Nuclear Corporation are noted. Article has interest for scientists and the general public.

306. TMI-1 RESTART
"TMI-1 Restart Activities Continue," E.G. Silver. Nuclear Safety 25 (March/April 1984): 275-277.
Class No.: E1.93:25-2. Item No.: 1051-H.
Series/Report No.: Order No. 84-079. OCLC: 1760906
Article, of special interest for the general public, focuses on legal and regulatory problems concerned with TMI-1 restart. Information about the November 1984 Rickover Report and the NRC recommendation endorsing a limited restart of TMI-1 is given.

307. TMI-2 ACCIDENT RECOVERY ACTIVITIES
"TMI-2 Recovery Progress," E.G. Silver. Nuclear Safety 25 (March/April 1984): 272-275.
Class No.: E1.93:25-2. Item No.: 1051-H.
Series/Report No.: Order No. 84-079. OCLC: 1760906

Journal Articles 117

Article focuses on testing of the polar crane and other activities designed to lead to the eventual defueling of the TMI-2 reactor. Allegations are noted concerning safety and management problems with TMI-2 clean-up. Article has interest for general public and scientists.

308. TMI-2 ACCIDENT RECOVERY ACTIVITIES
"DOE to Transport TMI Core to Idaho for Safety Check."
DOE This Month* 7 (April 1984): 1.
Class No.: E1.54:7-3. Item No.: 429-T-37.
Series/Report No.: Order No. 84-245. OCLC: 10434778
Brief general public interest report describes plans to remove damaged TMI-2 reactor core samples from Pennsylvania to Idaho for further testing.

309. TMI-2 ACCIDENT RECOVERY ACTIVITIES
"Japanese to Participate in TMI Research Under $18 Million, 5-Year Agreement." DOE This Month 7 (May 1984): 3.
Class No.: E1.54:7-5. Item No.: 429-T-37.
Series/Report No.: Order No. 84-324. OCLC: 10434778
Brief public interest report describes agreement, negotiated by U.S. Department of Energy representatives, whereby consortium of private Japanese firms agrees to contribute funds for research related to TMI-2 cleanup in return for access to TMI-2 research findings.

310. NUCLEAR POWER PLANT SAFETY
"Hydrogen Combustion and Control in Nuclear Reactor Containment Buildings," Loren Thompson [et al.]. Nuclear Safety 25 (May/June 1984): 350-372. Bibl.
Class No.: E1.93:25-3. Item No.: 1051-H.
Series/Report No.: Order No. 84-079. OCLC: 1760906
Technical report describes research program dealing with probable hydrogen behavior during a reactor core uncovery accident, emphasizing the TMI-2 accident findings. Report has special interest for engineers and scientists.

311. NUCLEAR POWER PLANT SAFETY
"Technical Note: LWR Safety after TMI," Doan L. Phung.
Nuclear Safety 25 (May/June 1984): 317-323. Bibl.
Class No.: E1.93:25-3. Item No.: 1051-H.
Series/Report No.: Order No. 84-079. OCLC: 1760906
Report, of interest to the public and scientists, concludes that U.S. light water reactors are considerably safer since the TMI-2 accident, because of improved equipment, new operating procedures, and improved training for plant operators. Author suggests chances of serious nuclear accidents will increase as more reactors come into operation.

*Note: DOE This Month continues Energy Insider beginning with Feb. 1984 issue.

312. NUCLEAR POWER PLANT SAFETY
"Technical Note: Reactor Accident Assessment from State Perspective," William P. Dornsife. Nuclear Safety 25 (May/June 1984): 323-328. Bibl.
Class No.: E1.93:25-3. Item No.: 1051-H.
Series/Report No.: Order No. 84-079. OCLC: 1760906
Based on TMI-2 experiences, this report recommends various protective actions to be taken during any future nuclear power plant accidents in Pennsylvania. Report has particular interest for engineers, scientists, and government planners.

313. TMI-1 RESTART
"TMI-1 Restart Still Enmeshed in Delays," E.G. Silver. Nuclear Safety 25 (May/June 1984): 414-416.
Class No.: E1.93:25-3. Item No.: 1051-H.
Series/Report No.: Order No. 84-079. OCLC: 1760906
Brief description of equipment problems and management issues which appear to be delaying TMI-1's restart. Report has general interest for the public and government officials.

314. TMI-2 ACCIDENT RECOVERY ACTIVITIES
TMI-2: "Action on Many Fronts," E.G. Silver. Nuclear Safety 25 (May/June 1984): 412-414.
Class No.: E1.93:25-3. Item No.: 1051-H.
Series/Report No.: Order No. 84-079. OCLC: 1760906
Report of general public interest, describes various TMI-2 clean-up activities, emphasizing both NRC and GPU Nuclear Corporation problems and activities related to TMI-2 clean-up in early 1984.

315. NUCLEAR POWER PLANT SAFETY
"Computerized Operator Decision Aids," A.B. Long. Nuclear Safety 25 (July/August 1984): 512-524. Bibl.
Class No.: E1.93:25-4. Item No.: 1051-H.
Series/Report No.: Order No. 84-079. OCLC: 1760906
Technical report, of special interest for engineers and scientists, evaluates the probably benefits of computer use in nuclear power plants to aid plant operators' decision making. Special attention is given to TMI-2 accident.

316. TMI-2 LEGAL PROBLEMS
"Metropolitan Edison Company Pleads Guilty." Nuclear Safety 25 (July/August 1984): 568.
Class No.: E1.93:25-4. Item No.: 1051-H.
Series/Report No.: Order No. 84-079. OCLC: 1760906
Brief report describes circumstances of guilty plea made by Met-Ed concerning falsifying and covering up TMI-2 reactor leak-rate records. The company's subsequent agreement to pay $1 million for TMI area emergency planning fund is noted. Article has general interest for public.

317. TMI-1 RESTART
"TMI-1 Restart Activities Continue," M.D. Muhlheim and E.G.

Silver. Nuclear Safety 25 (July/August 1984): 569-570.
Class No.: E1.93:25-4. Item No.: 1051-H.
Series/Report No.: Order No. 84-079. OCLC: 1760906
Report of general public interest summarizes various legal and regulatory activities in 1984, emphasizing Rickover report recommending restart of TMI-1.

318. TMI-2 ACCIDENT RECOVERY ACTIVITIES
"TMI-2 Recovery Progress," M.D. Muhlheim and E.G. Silver. Nuclear Safety 25 (July/August 1984): 568-569.
Class No.: E1.93:25-4. Item No.: 1051-H.
Series/Report No.: Order No. 84-079. OCLC: 1760906
Brief article, with interest for the general public, outlines TMI-2 clean-up activities completed by May 1984, including collection of second set of TMI-2 core samples and continued tests of the polar crane.

319. TMI-2 ACCIDENT RECOVERY ACTIVITIES
"TMI-2 Cleanup Activities." Nuclear Safety 25 (September/October 1984): 714.
Class No.: E1.93:25-5. Item No.: 1051-H.
Series/Report No.: Order No. 84-079. OCLC: 1760906
Report presents brief summary of 1984 clean-up problems at TMI-2 involving financial and legal problems, and notes clean-up progress. Article has general public interest.

320. NUCLEAR POWER PLANT REGULATION
"Technical Note: From Inprudence to Accident--the IEC Road to TMI," Frioyes Reisch. Nuclear Safety 25 (September/October 1984): 652-653.
Class No.: E1.93:25-5. Item No.: 1051-H.
Series/Report No.: Order No. 84-079. OCLC: 1760906
Author argues that International Electrotechnical Commission needs to take a stronger position on monitoring of nuclear power plants to ensure worldwide nuclear power plant safety in light of TMI-2 and other accidents. Opinions of author have general interest to the public.

321. TMI-1 RESTART
"TMI-1 Restart Activities Continue," M.D. Muhlheim and E.G. Shaw. Nuclear Safety 25 (September/October 1984): 713-714.
Class No.: E1.93:25-5. Item No.: 1051-H.
Series/Report No.: Order No. 84-079. OCLC: 1760906
Brief summary of activities related to proposed TMI-1 restart. Report has interest for the general public.

322. TMI-2 PSYCHOLOGICAL EFFECTS OF ACCIDENT
"Grantees Report at APA Meeting," Judy Folkenberg. ADAMHA News 10 (October 1984): 11+.
Class No.: HE20.8013:10-10. Item No.: 497-D-6.
Series/Report No.: Order No. 84-299. OCLC: 6416977

Survey suggests TMI-2 accident demoralized the general population of the TMI area and was particularly stressful for teenagers and mothers of preschool children. Some TMI workers experienced prolonged stress-related problems. Report has interest for the general public and health professionals.

323. TMI-2 ACCIDENT RECOVERY ACTIVITIES
"Hydrogen Transport in Containments: A Survey of Analytical Tools and Benchmark Experiments," Vincent P. Manno and Michael W. Golay. Nuclear Safety 25 (November/December 1984): 797-814. Bibl.
Class No.: E1.93:25-6. Item No.: 1051-H.
Series/Report No.: Order No. 84-079. OCLC: 1760906
Report reviews appropriate research studies of hydrogen safety in light water reactor containments, with special attention to the lessons of the TMI-2 emergency. Article has particular interest for engineers and scientists.

324. NUCLEAR POWER PLANT REGULATION
"Refueling for the Future--the American Scene," Nunzio J. Palladino. Nuclear Safety 25 (November/December 1984): 858-861. Bibl.
Class No.: E1.93:25-6. Item No.: 1051-H.
Series/Report No.: Order No. 84-079. OCLC: 1760906
Summary of speech by NRC chairman gives detailed immediate and long-term goals for U.S. nuclear power industry in light of TMI-2 accident findings. Report has general public interest.

325. TMI-1 RESTART
"TMI-1 Restart Schedule Clouded." Nuclear Safety 25 (November/December 1984): 848-849.
Class No.: E1.93:25-6. Item No.: 1051-H.
Series/Report No.: Order No. 84-079. OCLC: 1760906
Summarizes issues contributing to delays in restart of TMI-1, with particular attention to Governor Thornburgh's request for further NRC hearings and charges concerning TMI-1 management integrity. Report has general public interest.

326. TMI-2 ACCIDENT RECOVERY ACTIVITIES
"Three Mile 2 Head Removal Accomplished." Nuclear Safety 25 (November/December 1984): 845-846.
Class No.: E1.93:25-6. Item No.: 1051-H.
Series/Report No.: Order No. 84-079. OCLC: 1760906
Brief report, with general public interest, announces successful removal of damaged TMI-2 reactor vessel and projected plans designed to complete TMI-2 clean-up by mid-1988 or 1989.

327. TMI-1 RESTART
"Further Delay in TMI-1 Restart Decision," E.G. Silver. Nuclear Safety 26 (January/February 1985): 104-105.
Class No.: E1.93:26-1. Item No.: 1051-H.

Series/Report No.: Order No. 85-079. OCLC: 1760906
Report, of general public interest, describes NRC review of various management issues related to proposed TMI-1 restart.

328. NUCLEAR POWER PLANT SAFETY
"Nuclear Reactor Safety Goals and Assessment Principles," Donald J. Higson. Nuclear Safety 26 (January/February 1985): 1-13. Bibl.
Class No.: E1.93:26-1. Item No.: 1051-H.
Series/Report No.: Order No. 85-079. OCLC: 1760906
Technical report, with particular interest for nuclear engineers and scientists, compares and contrasts reactor safety in the United Kingdom and the U.S., with particular attention to the TMI-2 accident.

329. NUCLEAR POWER PLANT REGULATION
"Revised Licensee Event Report System," Gary T. Mays and Willis P. Porri. Nuclear Safety 26 (January/February 1985): 75-84. Bibl.
Class No.: E1.93:26-1. Item No.: 1051-H.
Series/Report No.: Order No. 85-079. OCLC: 1760906
Technical report describes forms for reporting light water reactor problems to the NRC by commercial nuclear power plants, following TMI-2 accident investigation recommendations. Report is directed to nuclear engineers.

330. TMI-2 ACCIDENT RECOVERY ACTIVITIES
"Financing of TMI-2 Cleanup," M.D. Muhlheim and E.G. Silver. Nuclear Safety 26 (March/April 1985): 237-238.
Class No.: E1.93:26-2. Item No.: 1051-H.
Series/Report No.: Order No. 85-079. OCLC: 1760906
Article, of general public interest, summarizes GPU Nuclear's problems in securing funds to complete proposed TMI-2 clean-up operations.

331. TMI-2, PSYCHOLOGICAL EFFECTS OF ACCIDENT
"Stress and the Emergency Worker," Marilyn Sargent. ADAMHA News 11 (April 1985): 8.
Class No.: HE20.8013:11-4. Item No.: 497-D-6.
Series/Report No.: Order No. 85-299. OCLC: 6416977
Research suggests persons exposed to technological disasters like the TMI-2 emergency suffer more chronic stress than those who experience a natural disaster, such as a flood or earthquake. Report has general interest for the public and psychologists.

332. TMI-2 ACCIDENT RECOVERY ACTIVITIES
"Fiscal Year 1986 DOE and NRC Administration Budget Proposals," E.G. Silver. Nuclear Safety 26 (May/June 1985): 398-402.
Class No.: E1.93:26-3. Item No.: 1051-H.
Series/Report No.: Order No. 85-079. OCLC: 1760906
Summarizes fiscal U.S. 1986 budget proposals, with special

attention to monies for light water reactor research and programs connected with TMI-2 clean-up activities.

333. TMI-1 RESTART
"Removal or Refusal of TMI-1 Restart: ASLB Chairman Becomes Issue," E.G. Silver. Nuclear Safety 26 (May/June 1985): 386-387.
Class No.: E1.93:26-3. Item No.: 1051-H.
Series/Report No.: Order No. 85-079. OCLC: 1760906
Brief article, of general interest to public, summarizes disagreements concerning activities of Chairman Smith, member of the Atomic Safety and Licensing Board, with regard to TMI-1 restart. No further public hearing on TMI-1 restart is scheduled.

334. TMI-2 HEALTH STUDIES
"Coping with the Threat of Radiation Exposure," Patricia Rikli [et al.]. Medical Bulletin of the U.S. Army, Europe 42 (June 1985): 21-23. Bibl.
Class No.: D101.42/3:42-6. Item No.: 325-B-04.
Series/Report No.: Order No. 85-338. OCLC: 2483943
In the light of health-related research studies following the TMI-2 emergency, the authors suggest probable health problems related to a serious nuclear accident. Report has particular interest for scientists and health professionals.

335. TMI-2 ACCIDENT RECOVERY ACTIVITIES
"Idaho's Loss-of-fluid Test Facility to Simulate TMI Core Damage." DOE This Month 8 (June, 1985): 1.
Class No.: E1.54:8-6. Item No.: 429-T-37.
Series/Report No.: Order No. 85-324. OCLC: 10434778
Brief report describes laboratory tests designed to simulate TMI-2 accident conditions so as to predict probable radioactive releases from the TMI-2 reactor core and fuel debris. Report has general interest for the public.

336. TMI-1 RESTART
"Three Mile Island 1 Restart Vote Again Likely," E.G. Silver. Nuclear Safety 26 (May/June 1985): 386.
Class No.: E1.93:26-3. Item No.: 1051-H.
Series/Report No.: Order No. 85-079. OCLC: 1760906
Brief report indicates NRC has decided that no further hearings will be required before a vote on proposed restart of TMI-1 is taken. Report has general interest for the public.

337. NUCLEAR POWER PLANT SAFETY
"Alternate Approaches to Nuclear Safety," Alan T. Carne. Nuclear Safety 26 (July/August 1985): 468-476. Bibl.
Class No.: E1.93:26-4. Item No.: 1051-H.
Series/Report No.: Order No. 85-079. OCLC: 1760906
Public opinion polls reflect a continuing lack of confidence in nuclear power plant safety following the TMI-2 accident.

Journal Articles 123

Author suggests public confidence in nuclear power can be restored only by the development of a safer reactor with little chance of meltdown. Three possible types of reactors are described. This technical report will have immediate interest for researchers, engineers, and the public.

338. TMI-2 ACCIDENT REPORTS
"Some Fuel Melted at TMI-2," E.G. Silver. Nuclear Safety 26 (July/August 1985): 528.
Class No.: E1.93:26-4. Item No.: 1051-H.
Series/Report No.: Order No. 85-079. OCLC: 1760906
Brief news item indicates that "a large amount of debris" visible on the damaged TMI-2's reactor floor was apparently "once molten." Remote camera pictures also indicate that significant parts of the reactor fuel reached 5080 degrees Fahrenheit, the melting point of uranium oxide. Report has interest for scientists and the public.

339. TMI-1 RESTART
"Three Mile Island 1 Restart Still Clouded," E.G. Silver. Nuclear Safety 26 (September/October 1985): 668-673.
Class No.: E1.93:26-5. Item No.: 1051-H.
Series/Report No.: Order No. 85-079. OCLC: 1760906
Report describes lengthy set of developments from June 1984 through May 1985 resulting in NRC decision to permit restart of TMI-1, subject to certain conditions mandated by the Commission and subsequent court appeals. Researchers, members of the legal profession and the public will have direct interest in chronological development of events pertaining to TMI-1 restart.

340. TMI-2 ACCIDENT RECOVERY ACTIVITIES
"TMI-2 Plenum Successfully Removed from Reactor." Nuclear Safety 26 (September/October 1985): 673.
Class No.: E1.93:26-5. Item No.: 1051-H.
Series/Report No.: Order No. 85-079. OCLC: 1760906
Brief news report from GPU Nuclear Corporation indicates successful removal of the plenum, housing the TMI-2 reactor's control-rod guide tubes, from the damaged reactor to an underwater storage area in the TMI-2 refueling canal. News item has direct interest for the public.

341. TMI-2 HEALTH STUDIES
"Longitudinal Analysis of Categorial Epidemiological Data: A Study of TMI," Stephen E. Fienberg [et al.]. Environmental Health Perspectives 36 (November 1985): 241-248. Bibl.
Class No.: HE20.3559:36-11. Item No.: 507-P-3.
Series/Report No.: Order No. 85-037. OCLC: 1727134
A technical evaluation of research methods used in longitudinal epidemiological study of the psychological well-being of mothers with young children who lived within ten miles of TMI during the 1979 TMI-2 emergency. Researchers, health professionals and scientists will have immediate interest in research method analysis.

PART 4

U.S. GOVERNMENT-SPONSORED PUBLICATIONS:
REPORTS*

342. INDEX, TMI-1 DOCUMENTS
FIND: Three Mile Island Nuclear Station, Unit 1. Prepared by James G. Smith for U.S. Atomic Energy Commission. Washington, D.C.: U.S. Atomic Energy Commission, 1971. 19p.
NTIS Order No.: FIND-502089(9-71). NTIS Prices: PC A03/MF A01.
Fiche listing of documents related to TMI's early development. See also entry no. 343.

343. INDEX, TMI-1 DOCUMENTS
FIND. Three Mile Island Nuclear Station, Unit 1. Revision 1. Prepared by James G. Smith for U.S. Atomic Energy Commission, Washington, D.C.: U.S. Atomic Energy Commission, 1972. 27p.
NTIS Order No.: FIND-50289(4-72). NTIS Prices: PC A03/MF A01.
Continuation of fiche listing of documents related to TMI-1's early development. See also entry no. 342.

344. TMI ENVIRONMENTAL IMPACT STATEMENT
Draft Environmental Impact Statement: the Environmental Considerations Related to the Proposed Issuance of an Operating License to Metropolitan Edison Co., Jersey Central Power & Light Co., and Pennsylvania Electric Co., for the TMI Nuclear Station Units 1 & 2, Docket Nos. 50-289 & 50-320. Prepared by U.S. AEC. Washington, D.C.: U.S. Atomic Energy Commission, 1972. 141p.
NTIS Order No.: EIS-PA-72-4772-D. NTIS Prices: PC E06.
Environmental considerations relative to the proposed construction and operation of two pressurized light water reactor units at TMI. Each proposed unit has two cooling towers to permit removal of waste heat from the closed-cycle cooling system.

345. INDEX, TMI-1 DOCUMENTS
FIND: Three Mile Island Nuclear Station, Unit 1. Prepared by H.D. Raleigh for U.S. Atomic Energy Commission. Washington, D.C.: U.S. Atomic Energy Commission, 1973. 28p. Bibl.

*First line of each entry consists of entry numbers and subject of the entry. Title and bibliographic details follow on subsequent lines.

Reports 125

NTIS Order No.: FIND-50289-R2. NTIS Prices: PC A03/MF A01.
Documents related to the early development of the TMI-1 reactor are listed and annotated.

346. TMI-1, OPERATING LICENSE
Three Mile Island Nuclear Station, Unit 1. Safety Evaluation Report. Prepared by Directorate of Licensing, U.S. Atomic Energy Commission. Washington, D.C.: U.S. Atomic Energy Commission, 1973. 222p. Bibl.
NTIS Order No.: Docket-50289-109. NTIS Prices: PC E08/MF A01.
The U.S. Atomic Energy Commission, forerunner of the U.S. Nuclear Regulatory Commission, presents technical review of Metropolitan Edison's application for operating license for TMI-1. Topics covered include site selection, environmental considerations, nuclear design, safety and health protection, radioactive waste disposal, emergency planning, etc. The AEC approved the TMI-1 operating license, subject to correction of specific items.

347. TMI-1, ENVIRONMENTAL IMPACT
An Evaluation of Environmental Data Relating to Select Nuclear Power Plant Sites: the TMI Nuclear Station Site. Prepared by Edwin D. Pentecost and Ishwar P. Muraka for U.S. Energy Research Development Administration. Springfield, Va.: NTIS, U.S. Dept. of Commerce, 1976. 13p.
NTIS Order No.: ANL/EIS-4. NTIS Prices: PC A02/MF A01.
OCLC: 7964465
TMI Unit 1 began operation in 1974. Analysis of environmental monitoring data, collected 1973 and 1974, indicates no adverse environmental effects to local plant or biotic life at the TMI-1 site or in the immediate area from the nuclear power plant's initial operations.

348. TMI-2, ENVIRONMENTAL EFFECTS
Operation of TMI Nuclear Station, Unit 2, Metropolitan Edison Co., Jersey Central Power and Light Co., Pennsylvania Electric Co., Docket No. 50-320. Final Supplement to Final Environmental Statement. Prepared by U.S. Nuclear Regulatory Commission. Washington, D.C.: U.S. Nuclear Regulatory Commission, 1976. 312p. Bibl.
NTIS Order No.: PB-261 501/1. NTIS Prices: PC A14/MF A01.
Supplementary information for TMI-2's environmental impact statement suggests the operation of TMI-2 will not cause any serious environmental problems. NRC report approves operating license application for TMI-2, subject to improvement of certain items and continuance of environmental monitoring activities.

349. TMI-2 SAFETY EVALUATION
Safety Evaluation Report: Operation of the Three Mile Island Nuclear Station, Unit 2, Metropolitan Edison Company, Jersey Central Power & Light Company, Docket No. 50-320. Prepared by U.S. Nuclear Regulatory Commission. Washington, D.C.: U.S. Nuclear Regulatory Commission, 1976. 182p.
NTIS Order No.: PB-260 801/6. NTIS Prices: PC A09/MF A01.

Technical report reviews and evaluates Metropolitan Edison's application for a license to operate TMI-2. Report covers various topics: site selection, nuclear reactor design requirements, engineering safety, electrical power generation, emergency procedures planning, radiation discharges and effects, radioactive waste disposal and public health. NRC approved the Met. Ed. proposal, subject to correction of certain items.

350. TMI-1 RADIATION RELEASES
An Aerial Radiological Survey of the TMI Station Nuclear Power Plant (Goldsboro, Pa.). Prepared by A.E. Fritzsche for U.S. Dept. of Energy. Springfield, Va.: NTIS, U.S. Dept. of Commerce, 1977. 20p. Bibl.
NTIS Order No.: Egg-1183-1710. NTIS Prices: Pc A02/MF A01.
OCLC: 9872936
An aerial survey of the TMI facility and surrounding area was made between August 2, 1976 and August 4, 1976 to monitor possible radiation releases. Geological characteristics and vegetation types are noted. No adverse radiation releases were discovered.

351. TMI-2 RADIATION RELEASES
Archiving the Paper Fallout from Three Mile Island. Prepared by H.L. Fisher and V.D. Caswell for U.S. Dept. of Energy. Springfield, Va.: NTIS, U.S. Dept. of Commerce, 1979. 7p. Bibl.
NTIS Order No.: UCID-18243. NTIS Prices: PC A02/MF A01.
From March 28, 1979 to April 18, 1979, the Atmosphere Release Advisory Capability (ARAC) Center at the Lawrence Livermore Laboratory supplied the Nuclear Regulatory Commission and the Department of Energy with meteorological records and radioactive release data related to the TMI-2 emergency. This technical report describes an information system designed to make this vast accumulation of data readily available for further research or for comparison purposes in the event of any future serious nuclear emergency.

352. TMI-2 RADIATION RELEASES
TMI-2 Decay Power: LASL Fission-Product and Actinide Decay Power Calculations for the President's Commission on the Accident at Three Mile Island. Prepared by T.R. England and W.B. Wilson for U.S. Dept. of Energy. Springfield, Va.: NTIS, U.S. Dept. of Commerce, 1979. 217p. Bibl.
NTIS Order No.: La-8041-MS. NTIS Prices: PC A10/MF A01.
Technical report summarizes data, using graphs and tables to estimate the probable rate of decay of TMI-2 fission products and actinide isotopes. Report was compiled at the request of the President's Commission on the Accident at Three Mile Island.

353. TMI-2 ACCIDENT RECOVERY ACTIVITIES
Post-accident Cleanup of Radioactivity at Three Mile Island. Prepared by R.E. Brooksbank for U.S. Dept. of Energy. Springfield, Va.: NTIS, U.S. Dept. of Commerce, 1979. 27p. Bibl.
NTIS Order No.: DOE/TIC-11023. NTIS Prices: MF A01.

Report presents preliminary report planning for decontamination activities at TMI-2, with special attention to problems of liquid waste processing and facilities needed to complete this important activity. Proposed removal of krypton gas from TMI-2's containment buildings is discussed.

354. TMI-2, FINANCIAL EFFECTS
Preliminary Analysis of the Effects of the Accident at Three Mile Island on Metropolitan Edison Company. Prepared by Temple, Barker and Sloane, Inc. for U.S. Dept. of Energy. Springfield, Va.: NTIS, U.S. Dept. of Commerce, 1979. 46p. Bibl.
NTIS Order No.: DE83006020. NTIS Prices: MF A01.

Report summarizes the immediate effects of the TMI-2 accident for Metropolitan-Edison's customers and the financial problems of Metropolitan-Edison and its parent corporation, General Public Utilities, Inc., caused by the TMI-2 accident.

355. TMI-2 RADIATION RELEASES
Simulation of the Three Mile Island Transient in Semiscale. Prepared by T.K. Larson [et al.] for U.S. Dept. of Energy. Springfield, Va.: NTIS, U.S. Dept. of Commerce, 1979. 94p. Bibl.
NTIS Order No.: EGG/SEMI-TR-010. NTIS Prices: PC A05/MF A01.

Report summarizes preliminary test data analysis findings obtained from eight simulations of the TMI-2 accident on March 28, 1979, using the Semiscale Mod-3 System to simulate the accident conditions.

356. TMI-2 ECONOMIC EFFECTS
Advertising/Public Relations Campaign to Combat the Negative Economic Impact Caused by the Nuclear Mishap at Three Mile Island, Pennsylvania. Prepared by Ketchum, MacLeod and Grove, Inc. for U.S. Economic Development Administration and Pennsylvania Dept. of Commerce. Springfield, Va.: NTIS, U.S. Dept. of Commerce, 1980. 25p. Bibl.
NTIS Order No.: PB81-104549. NTIS Prices: PC A02/MF A01.

Report discusses 1980 campaign initiated by Pennsylvania Department of Commerce utilizing radio and television advertising to offset the negative image of Pennsylvania caused by the TMI-2 accident. Advertising campaign tried to present a friendly image of Pennsylvania and to attract tourists to the state following the TMI-2 emergency.

357. TMI-2 ACCIDENT REPORTS
Crisis Contained, The DOE at Three Mile Island: a History. Prepared by Philip L. Cantelon and Robert C. Williams for U.S. Dept. of Energy. Springfield, Va.: NTIS, U.S. Dept. of Commerce, 1980. 243p. Bibl.
NTIS Order No.: DOE/EV/10278-TI. NTIS Prices: PC A11/MF A01.
OCLC: 7259200

Development of TMI-2 accident is described, focusing on the support activities taken by the DOE. Report is directed to engineers, scientists, and researchers. Note: This report was subsequently published as a book by Southern Illinois University Press in 1982.

358. TMI-2 HEALTH STUDIES
Estimates of Dose Due to Noble Gas Releases from the TMI Incident Using AIRDOS-EPA Computer Code. Prepared by S.J. Cotter [et al.] for U.S. Dept. of Energy. Springfield, Va.: NTIS, U.S. Dept. of Commerce, 1980. 57p. Bibl.
NTIS Order No.: ORNL-5649. NTIS Prices: PC A04/MF A01.
Technical report suggests that radiation exposure, for population living 55 miles or less from TMI during the TMI-2 emergency, was too small to produce harmful, long-term health problems. Engineers, scientists, and health professionals will find useful information in the report.

359. TMI-2 EVACUATION PLANNING
Evacuation Planning in the TMI Accident. Prepared by William W. Chenault [et al.] for Federal Emergency Management Agency. Washington, D.C.: Federal Emergency Management Agency, 1980. 210p. Bibl.
NTIS Order No.: AD-A080 104/3. NTIS Prices: PC A10/MF A01.
OCLC: 9047320
Report summarizes emergency plans for possible evacuation of population living in TMI vicinity during the TMI-2 emergency. Civil defense organizations and researchers will find this report most useful.

360. TMI-2 ACCIDENT RECOVERY ACTIVITIES
GEND Planning Report. Prepared by EG and G Idaho, Inc. for U.S. Dept. of Energy. Springfield, Va.: NTIS, U.S. Dept. of Commerce, 1980. 716p. Bibl.
NTIS Order No.: DE81025533. NTIS Prices: PC A99/MF A01.
Presents detailed planning for various TMI-2 recovery activities, including testing of all types of electrical instruments and other equipment, handling and removal of all TMI-2 radioactive waste materials, removal and examination of the damaged TMI-2 reactor core, etc. Technical report has particular interest for engineers, scientists, and researchers.

361. TMI-2 RADIATION RELEASES
Preliminary Comparisons Between Measurements and Model Calculations for the TMI Venting of Exp 85 Kr. Prepared by M.H. Dickerson for U.S. Dept. of Energy. Springfield, Va.: NTIS, U.S. Dept. of Commerce, 1980. 20p. Bibl.
NTIS Order No.: UCID. NTIS Prices: PC A02/MF A01.
Technical report reviews venting of Krypton-85 from the damaged TMI-2 reactor into the atmosphere through extensive radiation monitoring of air samples from various mobile air samplers and DOE helicopter measurements. Report has direct interest for DOE and researchers.

362. TMI-2 HEALTH STUDIES
Radiation Release and Health Effects Lessons from the TMI Accident. Application to Emergency Response Planning. Prepared

by F.R. Mynatt and C.D. Berger for U.S. Dept. of Energy. Springfield, Va.: NTIS, U.S. Dept. of Commerce, 1980. 6p. Bibl.
NTIS Order No.: DOE/TIC-11356. NTIS Prices: PC A02/MF A01.
An estimate of probable health effects of TMI-2 accident is presented in conjunction with emergency planning recommendations for U.S. nuclear power facilities under construction. Report has particular interest for emergency planners and health professionals.

363. TMI-2 HEALTH STUDIES
Radiation Release and Health Effects Lessons from the TMI Incident: Assessment of Objective Risks for Emergency Preparedness Planning. Prepared by C.D. Berger anf F.R. Mynatt for U.S. Dept. of Energy. Springfield, Va.: NTIS, U.S. Dept. of Commerce, 1980. 5p. Bibl.
NTIS Order No.: DOE/TIC-11360. NTIS Prices: PC A02/MF A01.
Radiation effects of TMI-2 accident are discussed in conjunction with low-level nuclear facility emergencies and serious emergencies requiring large-scale population evacuations. Emergency planners and health professionals will find this report useful.

364. TMI-2 ACCIDENT REPORTS
Review of Literature on the TMI Accident and Correlation to the LWR Safety Technology Program. Prepared by W.J. Miller for U.S. Dept. of Energy. Springfield, Va.: NTIS, U.S. Dept. of Commerce, 1980. 220p. Bibl.
NTIS Order No.: ALO-85. NTIS Prices: PC A10/MF A01.
Report reviews publicly available materials related to 1979 TMI-2 emergency with special correlation to safety plan for all U.S. commercial nuclear power facilities being prepared for U.S. Dept. of Energy. Report is directed to U.S. Dept. of Energy and has direct interest for nuclear engineers, scientists, and researchers.

365. TMI-2 ACCIDENT RECOVERY ACTIVITIES
Scoping Studies of the Alternative Options for Defueling, Packaging, Shipping, and Disposing of the TMI-2 Spent Fuel Core. Prepared by Allied-General Nuclear Services for U.S. Dept. of Energy. Springfield, Va.: NTIS, U.S. Dept. of Commerce, 1980. 217p. Bibl.
NTIS Order No.: AGNS-35900-1.5-79. NTIS Prices: PC A10/MF A01.
Technical report presents alternative plans designed to ship fuel safely from the damaged TMI-2 reactor and to permit subsequent examination, storage and eventual disposal of this fuel. Report is directed to U.S. Dept. of Energy and has immediate interest for engineers and others working on TMI-2 decontamination activities.

366. NUCLEAR POWER PLANT SAFETY
Some Implications of the TMI Accident for LMFBR Safety and Licensing: the Design Basis Issue. Prepared by Kenneth A. Solomon for U.S. Dept. of Energy [et al.]. Santa Monica, Calif.: Rand Corporation, 1980. 48p. Bibl.

NTIS Order No.: DE82009276. NTIS Prices: PC A03/MF A01.
OCLC: 7118825
   Technical report outlines recommendations to improve the safety operations of light water reactors (LWR's) and to affect the design of a liquid metal fast breeder reactor (LMFBR) following TMI-2 accident findings. Report has direct use for engineers, scientists, etc. seeking to improve nuclear reactor safety in U.S. commercial nuclear power facilities.

367. TMI-2 RADIATION STUDIES
   Technical Assessment of Radiation Overexposure at TMI from August 28, 1979 Entry into the Unit 2 Fuel-Handling Building Make-up Valve Room. Prepared by B.L. Rich and S.R. Adams for U.S. Dept. of Energy. Springfield, Va.: U.S. Dept. of Commerce, 1980. 28p. Bibl.
NTIS Order No.: DE82004040. NTIS Prices: PC A03/MF A01.
   Research report confirms findings of earlier Metropolitan Edison report that six authorized personnel were not subjected to hazardous radiation exposure during August 28, 1979 entry of TMI-2's Fuel Handling Building. Report has particular use for engineers, scientists, and other planning TMI-2 decontamination activities.

368. TMI-2 ACCIDENT REPORTS
   Three Mile Island: A Year Later. Prepared by D.B. Trauger for U.S. Dept. of Energy. Springfield, Va.: NTIS, U.S. Dept. of Commerce, 1980. 8p. Bibl.
NTIS Order No.: DOE/TIC-11172. NTIS Prices: PC A02/MF A01.
   Brief report is presented on TMI-2 accident, with attention to nuclear reactor new safety procedures, reactor licensing changes and public opinion concerning nuclear power plant safety.

369. TMI-2 ACCIDENT RECOVERY ACTIVITIES
   Three Mile Island: Then and Now. Prepared by D.B. Trauger for U.S. Dept. of Energy. Springfield, Va.: NTIS, U.S. Dept. of Commerce, 1980. 9p. Bibl.
NTIS Order No.: DOE/TIC-11235. NTIS Prices: PC A02/MF A01.
   TMI-2 accident is described, with attention to ongoing decontamination activities at TMI-2. Report is directed to U.S. Dept. of Energy. Technical report summary is useful for researchers.

370. TMI-2 ACCIDENT REPORTS
   A Prediction of TMI-2 Core Temperatures from the Fission Product Release History. Prepared by J. Rest and C.E. Johnson for U.S. Dept. of Energy. Springfield, Va.: NTIS, U.S. Dept. of Commerce, 1980. 23p. Bibl.
NTIS Order No.: DE82004492. NTIS Prices: PC A02/MF A01.
OCLC: 9963916
   Technical report discusses probable fission product releases during the TMI-2 accident. Available data indicate that rapid fuel temperature increases during the TMI-2 emergency led to extensive fission product releases and fuel fracturing. A companion volume to

Reports 131

this work is A Code for Predicting the Temperature and Oxidation of Undercooled Cores, prepared by P. B. Abramson and others.

371. TMI-2 ACCIDENT RECOVERY ACTIVITIES
TMI-2 Decay Power: LASL Fission-Product and Actinide Decay Power Calculations for the President's Commission on the Accident at Three Mile Island. Prepared by T.R. England and W.B. Wilson for U.S. Dept. of Energy. Springfield, Va.: NTIS, U.S. Dept. of Commerce, 1980. 228p. Bibl.
NTIS Order No.: LA-8041-MS(REV). NTIS Prices: PC A11/MF A01.

A series of letters summarizes data compiled regarding the probable decay rate of TMI-2 radioactive waste materials and suggests various alternatives for decontamination of these fission products.

372. TMI-2 RADIATION STUDIES
Utilization of the Atmospheric Release Advisory Capability (ARAC) Services During and After the TMI Accident. Prepared by J.B. Knox [et al.] for U.S. Dept. of Energy. Springfield, Va.: U.S. Dept. of Commerce, 1980. 37p. Bibl.
NTIS Order No.: UCRL-52959. NTIS Prices: PC A03/MF A01.

Technical report describes ARAC's radiation monitoring of atmosphere in TMI vicinity during 1979 emergency and presents subsequent radiation exposure calculations for TMI area residents. Report had use for President's Commission on TMI, the NRC, the U.S. Dept. of Energy and researchers.

373. TMI-2 ACCIDENT RECOVERY ACTIVITIES
Accountability Study for TMI-2 Fuel. Prepared by P. Goris and D.D. Scott for U.S. Dept. of Energy. Springfield, Va.: NTIS, U.S. Dept of Commerce, 1981. 145p. Bibl.
NTIS Order No.: GEND-016. NTIS Prices: PC A07/MF A01.

Technical report discusses problems of identifying, measuring and inventorying the TMI-2 fuel as it is removed from the TMI-2 reactor and during subsequent examinations. Report is directed to U.S. Dept. of Energy and has particular use for engineers and others engaged in clean-up of TMI-2 reactor.

374. TMI-2 ACCIDENT RECOVERY ACTIVITIES
Canister-design Considerations for Packaging of TMI Unit 2 Damaged Fuel and Debris. Prepared by G.A. Townes for U.S. Dept. of Energy. Springfield, Va.: NTIS, U.S. Dept. of Commerce, 1981. 91p. Bibl.
NTIS Order No.: DE82000885. NTIS Prices: PC A05/MF A01.

Technical report presents proposed design for a canister to be used for storage of damaged TMI-2 fuel and which may be transported and subsequently reprocessed or buried. Report has particular use for engineers and others engaged in clean-up of TMI-2 reactor.

375. TMI-2 RADIATION RELEASES
Characterization of an Aerosol Sample from TMI Reactor Auxiliary

Building. Prepared by G.M. Kanapilly [et al.] for U.S. Dept. of Energy. Springfield, Va.: U.S. Dept. of Commerce, 1981. 13p. NTIS Order No.: LMF-70. NTIS Prices: PC A02/MF A01. OCLC: 9908941
Technical report describes collection of air samples from TMI-2 auxiliary building following the TMI-2 emergency, and subsequent tests to determine the samples' radioactive composition.

376. TMI-2 RADIATION STUDIES
Characterization of the TMI Unit 2 Reactor Building Atmosphere Prior to the Reactor Building Purge. Prepared by J.K. Hartwell [et al.] for U.S. Dept. of Energy. Springfield, Va.: NTIS, U.S. Dept. of Commerce, 1981. 48p. Bibl.
NTIS Order No.: GEND-005. NTIS Prices: PC A03/MF A01.
Report describes air sampling procedures and analyses of air samples taken from TMI-2 reactor containment building. Technical study was done for U.S. Dept. of Energy and had direct use for decontamination of TMI-2 containment building.

377. TMI-2 EMERGENCY EVACUATION PLANNING
Citizen Evacuation in Response to Nuclear and Nonnuclear Threats: Final Report. Prepared by Ronald W. Perry for U.S. Federal Emergency Management Agency. Seattle, Wash.: Battelle Human Affairs Research Ctrs., 1981. 105p. Bibl.
NTIS Order No.: DE82901237. NTIS Prices: PC A06/MF A01. OCLC: 8787823
Research study discusses possible civilian evacuation decision procedures and early warning devices for the serious nuclear emergency at TMI-2, three serious river floods and the eruption of a volcano. The study compares civilian defense activities for the various emergencies, public reaction to the various crises and implications for future evacuation planning.

378. TMI-2 RADIATION RELEASES
Citizen Radiation Monitoring Program for the TMI Area. Prepared by A.J. Baratta [et al.] for U.S. Dept. of Energy. Springfield, Va.: NTIS, U.S. Dept. of Commerce, 1981. 419p. Bibl.
NTIS Order No.: DE81025487. NTIS Prices: PC A18/MF A01.
Report describes program to train citizens living near nuclear power facilities to make radiation measurements and how to interpret these data. Program could be widely used in communities near U.S. commercial nuclear power facilities.

379. TMI-2 ACCIDENT REPORTS
A Code for Predicting the Temperature and Oxidation of Undercooled Cores. Prepared by P.B. Abramson [et al.] for U.S. Dept. of Energy. Springfield, Va.: NTIS, U.S. Dept. of Commerce, 1981. 65p. Bibl.
NTIS Order No.: DE82004493. NTIS Prices: PC A04/MF A01. OCLC: 9963535
TMI-2 accident simulations suggest the TMI-2 reactor was

severely damaged between 100 minutes and 174 minutes after the emergency began when the reactor core experienced loss of coolant circulation. Through use of a computer code, a method of analyzing temperature and oxygen changes during a serious nuclear power plant emergency like the TMI-2 emergency has been improved. A companion volume is A Prediction of TMI-2 Core Temperatures from the Fission Product Release History, prepared by J. Rest and C.E. Johnson.

380. TMI-2 RADIATION RELEASES
Examination Results of the TMI Radiation Detector HP-R-211.
Prepared by M.B. Murphy [et al.] for U.S. Dept. of Energy. Springfield, Va.: NTIS, U.S. Dept. of Commerce, 1981. 133p. Bibl.
NTIS Order No.: DE82000939. NTIS Prices: PC A07/MF A01.

Technical report records probable causes for failure of radiation monitor, probable amounts of radiation received by the detector during the TMI-2 accident and possible decontamination procedures. Report will have direct use for engineers and scientists engaged in TMI-2 clean-up activities.

381. TMI-2 ACCIDENT RECOVERY ACTIVITIES
Feasibility of Vitrifying EPICOR II Organic Resins. Prepared by J.L. Buelt for U.S. Dept. of Energy. Springfield, Va.: NTIS, U.S. Dept. of Commerce, 1981. 24p. Bibl.
NTIS Order No.: DE820003954. NTIS Prices: PC A02/MF A01.

Technical report describes feasibility studies which explore possibility of using a single-step incineration/vitrification process to accelerate radioactive waste processing at TMI-2. Report is directed to U.S. Dept. of Energy and should have direct use for engineers and others engaged in TMI-2 clean-up operations.

382. TMI-2 ACCIDENT RECOVERY ACTIVITIES
Field Measurements and Interpretation of TMI-2 Instrumentation: CF-1-PT3. Prepared by J.E. Jones [et al.] for U.S. Dept. of Energy. Springfield, Va.: NTIS, U.S. Dept. of Commerce, 1981. 42p. Bibl.
NTIS Order No.: DE82003587. NTIS Prices: MF A01.

Technical report indicates this pressure monitor on Core Flood Tank of TMI-2 continued to operate even though pressure signal was reduced. Technical analysis would have direct use for engineers, and others supervising TMI-2 clean-up activities.

383. TMI-2 ACCIDENT RECOVERY ACTIVITIES
Field Measurements and Interpretation of TMI-2 Instrumentation: CF-1-PT4. Prepared by J.E. Jones [et al.] for U.S. Dept. of Energy. Springfield, Va.: NTIS, U.S. Dept. of Commerce, 1981. 42p. Bibl.
NTIS Order No.: DE2003587. NTIS Prices: MF A01.

Tests suggest this pressure monitor on Core Flood Tank of TMI-2 continued to operate with a reduced pressure signal. Technical report would have direct use for engineers and others supervising TMI-2 clean-up activities.

384. TMI-2 ACCIDENT RECOVERY ACTIVITIES
Interim Status Report of the TMI Personnel-Dosimetry Project.
Prepared by B.L. Rich [et al.] for U.S. Dept. of Energy. Springfield, Va.: NTIS, U.S. Dept. of Commerce, 1981. 94p. Bibl.
NTIS Order No.: DE81025672. NTIS Prices: PC A05/MF A01.
Technical report evaluates various dosimeter systems, suitable for TMI-2 recovery operations, and recommends use of a new Idaho National Engineering Laboratory prototype system. Report will have direct use for planners of TMI-2 clean-up operations.

385. TMI-2 ACCIDENT RECOVERY ACTIVITIES
In-Vessel Inspection Before Head Removal: TMI-2: Phase 1.
Prepared by N.E. Calloway [et al.] for U.S. Dept. of Energy. Springfield, Va.: NTIS, U.S. Dept. of Commerce, 1981. 90p. Bibl.
NTIS Order No.: DE81029492. NTIS Prices: PC A05/MF A01.
Technical report discusses remote viewing equipment designed to assist with preliminary examination of TMI-2 reactor before TMI-2's reactor head is removed. Report will assist engineers and others planning for decontamination activities at TMI-2.

386. TMI-2 ACCIDENT RECOVERY ACTIVITIES
In-Vessel Inspection Before Head Removal: TMI-2: Phase 2.
Prepared by D.W. Greenlee for U.S. Dept. of Energy. Springfield, Va.: NTIS, U.S. Dept. of Commerce, 1981. 122p. Bibl.
NTIS Order No.: DE81026957. NTIS Prices: PC A06/MF A01.
Technical report continues discussion of remote viewing equipment to provide a view of the top part of TMI-2's reactor core. Report will assist with decontamination activities at TMI-2.

387. TMI-2 ACCIDENT RECOVERY ACTIVITIES
Investigation of Acoustic Methods for Detecting TMI-2 Fuel Debris. Prepared by L.S. Beller for U.S. Dept. of Energy. Springfield, Va.: NTIS, U.S. Dept. of Commerce, 1981. 45p. Bibl.
NTIS Order No.: DE82001005/XAB. NTIS Prices: PC A03/MF A01.
Technical report discusses experiments using ultrasmic measurements and an accompanying instrument system for locating and measuring TMI-2 fuel debris now located in the reactor's piping system. Report will be useful for engineers and others planning for decontamination of TMI-2.

388. TMI-2 RADIATION STUDIES
Measurements of exp 129 I and Radioactive Particulate Concentrations in the TMI-2 Containment Atmosphere During and After the Venting. Prepared by J.E. Cline [et al.] for U.S. Dept. of Energy. Springfield, Va.: NTIS, U.S. Dept. of Commerce, 1981. 25p. Bibl.
NTIS Order No.: GEND-009. NTIS Prices: PC A02/MF A01.
Technical report presents data on changes in concentration of Iodine 129 and Krypton 85 during venting of TMI-2 containment building between June 28, 1980 and July 11, 1980. Data obtained were useful in planning for further TMI-2 decontamination activities.

389. TMI-2 ACCIDENT RECOVERY ACTIVITIES
Neutronic Analysis of TMI Unit 2 Ex-core Detector Response.
Prepared by D.J. Malloy and Y.I. Chang for U.S. Dept. of Energy.
Springfield, Va.: NTIS, U.S. Dept of Commerce, 1981. 37p. Bibl.
NTIS Order No.: DE82004040. NTIS Prices: PC A03/MF A01.
Technical report presents analysis of ex-core detector response during TMI-2 accident, with possible explanations for abnormal behavior of the detector. Report would have use for engineers and others interested in accuracy of TMI-2 reactor instrumentation.

390. TMI-2 ACCIDENT RECOVERY ACTIVITIES
Nondestructive Techniques for Assaying Fuel Debris in Piping at TMI-Unit 2. Prepared by K. Vinjamuri [et al.] for U.S. Dept. of Energy. Springfield, Va.: NTIS, U.S. Dept. of Commerce, 1981. 159p. Bibl.
NTIS Order No.: DE82003039. NTIS Prices: PC A08/MF A01.
Technical report discusses various techniques most suitable for isolating fuel debris in primary coolant of the TMI-2 reactor. Report recommends use of passive gamma ray technique for TMI-2 piping. Report is directed to engineers and others planning TMI-2 clean-up operations.

391. PUBLIC OPINION, NUCLEAR POWER PLANT SAFETY
Nuclear Energy Risks and Benefits. Prepared by Steven D. Jansen [et al.] for Environmental Protection Agency. Springfield, Va.: NTIS, U.S. Dept. of Commerce, 1981. 115p. Bibl.
NTIS Order No.: PB82-109703. NTIS Prices: PC A06/MF A01.
Report developed as a part of the research program of the Ohio River Basin Energy Study covering Kentucky, West Virginia, and parts of Illinois, Indiana, Ohio, and Pennsylvania. Includes general information on risks and benefits of commercial use of nuclear power to generate electricity in these states. The TMI-2 emergency is discussed, with an evaluation of the lessons learned from the 1979 accident.

392. TMI-2 RADIATION STUDIES
Population Dose Estimation from a Hypothetical Release of 2.4×10^6 Curies of Noble Gases and 1×10^4 Curies of exp 131 I at TMI Nuclear Station, Unit 2. Prepared by C.D. Berger [et al.] for U.S. Dept. of Energy. Springfield, Va.: NTIS, U.S. Dept. of Commerce, 1981. 30p. Bibl.
NTIS Order No.: DE81030701. NTIS Prices: PC A03/MF A01.
As part of the study of the President's Commission on the Accident at TMI, environmental consequences of a high release iodine 131 was projected for human populations within a 50-mile radius of TMI-2. Health professionals and environmental planners have immediate interest in this study.

393. PUBLIC OPINION, NUCLEAR POWER PLANT SAFETY
Public Opinion and Nuclear Power, January 1970-October 1981. (Citations from the NTIS Data Base). Prepared by the National

Technical Information Service. Springfield, Va.: U.S. Dept. of
Commerce, 1981. 97p. Bibl.
NTIS Order No.: PB82-855636. NTIS Prices: PC N01/MF N01.
 This bibliography lists some seventy-three NTIS reports dealing with the public's view of nuclear power as a commercial source of electric power. Topics covered include choice of sites, environmental considerations, nuclear waste disposal, health and safety factors, and reaction to the TMI-2 emergency.

394. TMI-2 ACCIDENT RECOVERY ACTIVITIES
Preliminary Characterization of EPICOR II Prefilter 16 Liner.
Prepared by EG and G Idaho, Inc. for U.S. Dept. of Energy.
Springfield, Va.: NTIS, U.S. Dept. of Commerce, 1981. 35p.
Bibl.
NTIS Order No.: DE82003588. NTIS Prices: PC A03/MF A01.
 Technical report presents development of EPICOR prefilter liner, designed for use in decontaminating radioactive water in TMI-2 auxillary and fuel handling building, emphasizing shipment of liners, design of specialized handling equipment, problems, etc. Engineers and others in charge of TMI-2 clean-up activities will find this report most useful.

395. TMI-2 ACCIDENT RECOVERY ACTIVITIES
Proposed Design Requirements for High-Integrity Containers Used to Store, Transport, and Dispose of High-Specific-Activity, Low-Level Radioactive Wastes from TMI Unit 2. Prepared by M.G. Vigil [et al.] for U.S. Dept. of Energy. Springfield, Va.: NTIS, U.S. Dept. of Commerce, 1981. 50p. Bibl.
NTIS Order No.: SAND-81-0567. NTIS Prices: PC A03/MF A01.
 Technical report proposes design requirements for containers to be used for storage, transportation and/or disposal of radioactive resin wastes produced by the EPICOR II waste treatment system. Proposed container design has been developed to guarantee safe disposal of TMI-2 low-level radioactive wastes. Report is directed to nuclear engineers and scientists.

396. TMI-2 ACCIDENT RECOVERY ACTIVITIES
Quick Look Report, Entry 3: Three Mile Island Unit 2, October 16, 1980. Prepared by EG and G Idaho, Inc. for U.S. Dept. of Energy. Springfield, Va.: NTIS, U.S. Dept. of Commerce, 1981. 34p. Bibl.
NTIS Order No.: DE81028758. NTIS Prices: PC A03/MF A01.
 Report describes radiation monitoring and technical tasks accomplished by a five-man entry team during approximately 90 minutes in the TMI-2 containment building on October 16, 1980. These entries are useful for engineers and others in charge of planning TMI-2 decontamination activities.

397. TMI-2 ACCIDENT RECOVERY ACTIVITIES
Quick Look Report, Entry 4: Three Mile Island Unit 2, November 13, 1980. Prepared by G.E. Eidam for U.S. Dept. of Energy.

Reports
137

Springfield, Va.: NTIS, U.S. Dept. of Commerce, 1981. 32p. Bibl. NTIS Order No.: DE81023966. NTIS Prices: PC A03/MF A01.

Report describes various tasks performed during entry of TMI-2 containment buildings, including radiation monitoring, a decontamination test and photography and videotaping at different elevations. Report has direct use for engineers and others who are planning TMI-2 decontamination activities.

398. TMI-2 ACCIDENT REPORTS
Response of the SPND Measurement System to Temperature During the TMI Unit 2 Accident. Prepared by N. Wilde and J.L. Morrison, Jr. for U.S. Dept. of Energy. Springfield, Va.: NTIS, U.S. Dept. of Commerce, 1981. 88p. Bibl.
NTIS Order No.: DE82004663. NTIS Prices: PC A04/MF A01.

Technical report suggests probable causes for failure of Self Powered Neutron Detector (SPND) Measuring System to show fuel rod temperatures during the TMI-2 accident. Report will have direct use for engineers and others concerned with TMI-2's accident development.

399. TMI-2 ACCIDENT REPORTS
TMI-2 Accident: Postulated Heat Transfer Mechanisms and Available Data Base. Prepared by R. Viskanta and A.K. Mohanty for U.S. Nuclear Regulatory Commission. Springfield, Va.: NTIS, U.S. Dept. of Commerce, 1981. 119p. Bibl.
NTIS Order No.: NUREG/CR-2121. NTIS Prices: PC A06/MF A01.

In the light of the TMI-2 accident findings, this technical report identifies areas relating to light water reactor safety that require further research. Additional research is needed to improve knowledge of basic heat transfer and fluid friction processes during a light water reactor accident similar to the TMI-2 emergency. Computer codes are being developed to better predict heat transfer and fluid characteristics within a nuclear reactor.

400. TMI-2 RADIATION RELEASES
TMI Nuclear Reactor Accident of March 1979. Environmental Radiation Data. Volume I. A Report to the President's Commission on the Accident at TMI. Prepared by Erich W. Bretthauer [et al.]. Springfield, Va.: NTIS, U.S. Dept. of Commerce, 1981. 762p.
NTIS Order No.: PB81-175374. NTIS Prices: MF A01.

Includes environmental radiation data listings collected in vicinity of TMI following the March 1979 accident. Data were collected by EPA, DOE, HHS, NRC, the Commonwealth of PA, or Bethlehem Steel Corporation. For volumes II-VI and Update Volume see the next six documents. These data are most valuable in evaluating the environmental consequences of the TMI-2 accident.

401. TMI-2 RADIATION RELEASES
TMI Nuclear Reactor Accident of March 1979. Environmental Radiation Sata. Volume II. A Report to the President's Commission on the Accident at TMI. Prepared by Erich Bretthauer [et al.].

Springfield, Va.: NTIS, U.S. Dept. of Commerce, 1981. 578p.
NTIS Order No.: PB 81-175382. NTIS Prices: MF A01.
See Volume I (entry no. 400) for description.

402. TMI-2 RADIATION RELEASES
TMI Nuclear Reactor Accident of March 1979. Environmental Radiation Data. Volume III. A Report to the President's Commission on the Accident at TMI. Prepared by Erich W. Bretthauer [et al.].
Springfield, Va.: NTIS, U.S. Dept. of Commerce, 1981. 1347p.
NTIS Order No.: PB 81-175390. NTIS Prices: MF A01.
See Volume I (entry no. 400) for description.

403. TMI-2 RADIATION RELEASES
TMI Nuclear Reactor Accident of March 1979. Environmental Radiation Data. Volume IV. A Report to the President's Commission on the Accident at TMI. Prepared by Erich W. Bretthauer [et al.].
Springfield, Va.: NTIS, U.S. Dept. of Commerce, 1981. 385p.
NTIS Order No.: PB 81-175408. NTIS Prices: MF A01.
See Volume I (entry no. 400) for description.

404. TMI-2 RADIATION RELEASES
TMI Nuclear Reactor Accident of March 1979. Environmental Radiation Data. Volume V. A Report to the President's Commission on the Accident at TMI. Prepared by Erich W. Bretthauer [et al.].
Springfield, Va.: NTIS, U.S. Dept. of Commerce, 1981. 771p.
NTIS Order No.: PB 81-175416. NTIS Prices: MF A01.
See volume I (entry no. 400) for description.

405. TMI-2 RADIATION RELEASES
TMI Nuclear Reactor Accident of March 1979. Environmental Radiation Data. Volume VI. A Report to the President's Commission on the Accident at TMI. Prepared by Erich W. Bretthauer [et al.].
Springfield, Va.: NTIS, U.S. Dept. of Commerce, 1981. 158p.
NTIS Order No.: PB 81-175424. NTIS Prices: MF A01.
See volume I (entry No. 400) for description.

406. TMI-2 RADIATION RELEASES
TMI Nuclear Reactor Accident of March 1979. Environmental Radiation Data: Update. A Report to the President's Commission on the Accident at TMI. Prepared by Erich W. Bretthauer [et al.].
Springfield, Va.: NTIS, U.S. Dept. of Commerce, 1981. 333p.
NTIS Order No.: PB 81-175366. NTIS Prices: MF A01.
See volume I (entry no. 400) for description.

407. TMI-2 RADIATION RELEASES
TMI-2 Reactor Building Purge: Kr-85 Venting. Prepared by L.T. Kripps for U.S. Dept. of Energy. Springfield, Va.: NTIS, U.S. Dept. of Commerce, 1981. 176p. Bibl.
NTIS Order No.: GEND-013. NTIS Prices: PC A09/MF A01.
    Technical report presents an evaluation of the major activities involved in decontaminating TMI-2's containment building by venting

Krypton 85 into the atmosphere. Environmental planners, appropriate government agencies and those planning TMI-2 recovery activities will find this report useful.

408. TMI-2 ACCIDENT RECOVERY ACTIVITIES
Three Mile Island Unit-2 Core Status Summary: A Basis for Tool Development for Reactor Disassembly and Defueling. Prepared by D.W. Croucher for U.S. Dept. of Energy. Springfield, Va.: NTIS, U.S. Dept. of Commerce, 1981. 73p. Bibl.
NTIS Order No.: GEND-007. NTIS Prices: PC A04/MF A01.

Evaluates the probable core damage to the reactor caused by the TMI-2 accident, and describes the probable condition of the TMI-2 reactor core following the emergency. Report will have particular use for engineers and others planning TMI-2 recovery operations.

409. TMI-2 ACCIDENT RECOVERY ACTIVITIES
TMI (Three Mile Island) Unit-2 Technical Information and Examination Program Update. Prepared by EG and G Idaho, Inc. for U.S. Dept. of Energy. Springfield, Va.: NTIS, U.S. Dept. of Commerce, 1981. 12p. Bibl.
NTIS Order No.: DE82007358. NTIS Prices: PC A02/MF A01.

Report discusses various problems necessary for decontamination of TMI-2, including a submerged demineralizer system for contaminated water in TMI-2 containment buildings, development of a polar crane, information for entries of TMI-2 containment building, etc. A good summary of activities related to the TMI-2 recovery program.

410. TMI-2 ACCIDENT RECOVERY ACTIVITIES
Analysis Data on Samples from the TMI-2 Reactor-Coolant System 2nd Reactor-Coolant Bleed Tank. Prepared by R.L. Nitschke for U.S. Dept. of Energy. Springfield, Va.: NTIS, U.S. Dept. of Commerce, 1982. 31p. Bibl.
NTIS Order No.: DE82016227. NTIS Prices: PC A03/MF A01.

Evaluates methods of analysis and results of liquid samples tested for radionuclide concentrations taken from TMI-2 Reactor Coolant Bleed Tanks and TMI-2 reactor coolant, between March 29, 1979 and August 14, 1979. Test results would have direct use for engineers, etc. planning for TMI-2 decontamination activities.

411. TMI-2 ACCIDENT RECOVERY ACTIVITIES
Characterization of EPICOR II Prefilter Liner 16. Prepared by J.D. Yesso [et al.] for U.S. Dept. of Energy. Springfield, Va.: NTIS, U.S. Dept. of Commerce, 1982. 62p. Bibl.
NTIS Order No.: DE82022228. NTIS Prices: PC A04/MF A01.

Detailed analysis of sampling and analysis of contents of the liner, to determine the liner's usefulness in processing TMI-2 radioactive wastes. Technical report would be useful for persons with responsibility for planning TMI-2 decontamination activities.

412. TMI-2 RADIATION RELEASES
Comparison of Occupational Dose Estimates for TMI-2 Recovery

Operation. Prepared by G.R. Hoenes for U.S. Dept. of Energy.
Springfield, Va.: NTIS, U.S. Dept. of Commerce, 1982. 8p. Bibl.
NTIS Order No.: DE83015088. NTIS Prices: PC A02/MF A01.
   Suggests reasons for two different estimates of probable occupational radiation exposure for personnel who are expected to participate in TMI-2 decontamination activities. Technical report will have immediate use in planning TMI-2 decontamination activities.

413. TMI-2 ACCIDENT RECOVERY ACTIVITIES
   Controlled Air Incinerator Conceptual Design Study. Prepared by EG and G Idaho, Inc. for U.S. Dept. of Energy. Springfield, Va.: NTIS, U.S. Dept. of Commerce, 1982. 101p. Bibl.
NTIS Order No.: DE82010039/XAB.
   Presents design for a controlled air incinerator for use in disposing of low-level combustible waste materials from the TMI-2 facility. Cost estimates, packaging and shipment of these wastes are also included. Report will have direct use for persons planning TMI-2 decontamination activities and for environmental planners.

414. TMI-2 HYDROGEN BURN
   Estimated Temperatures of Organic Materials in the TMI-2 Reactor Building During Hydrogen Burn. Prepared by H.W. Schultz and P.K. Nagata for U.S. Dept. of Energy. Springfield, Va.: NTIS, U.S. Dept. of Commerce, 1982. 46p. Bibl.
NTIS Order No.: DE83005355. NTIS Prices: PC A03/MF A01.
   Technical report estimates extent of hydrogen burn during 1979 TMI-2 emergency, using photographs and material samples. Report would have use for persons planning TMI-2 decontamination activities.

415. TMI-2 ACCIDENT RECOVERY ACTIVITIES
   Development of a Prototype Gas Sampler for EPICOR II Prefilter liners. Prepared by J.M. Bower for U.S. Dept. of Energy. Springfield, Va.: NTIS, U.S. Dept. of Commerce, 1982. 30p. Bibl.
NTIS Order No.: DE83000758. NTIS Prices: PC A03/MF A01.
   Discusses criteria for testing equipment and necessary systems to sample gases and other contents from liners. Report has use for persons planning TMI-2 decontamination activities.

416. TMI-2 ACCIDENT RECOVERY ACTIVITIES
   Development of in Situ Test Procedures for TMI-2 Axial Power Shaping Rod-drive Mechanisms. Prepared by J.A. Gannon for U.S. Dept. of Energy. Springfield, Va.: NTIS, U.S. Dept. of Commerce, 1982. 48p. Bibl.
NTIS Order No.: DE83004395. NTIS Prices: PC E03/MF A01.
   Describes procedures designed to simulate TMI-2's reactor's control rods operation at low speeds and to assist in determining the condition of TMI-2's reactor core. Report will have use for engineers and others planning TMI-2's decontamination activities.

417. TMI-2 ACCIDENT RECOVERY ACTIVITIES
   Estimated Source Terms for Radionuclides and Suspended

Reports 141

Particulates During TMI-2 Defueling Operations. Report on Phase I. Prepared by P.G. Voilleque for U.S. Dept. of Energy. Springfield, Va.: NTIS, U.S. Dept. of Commerce, 1982. 26p. Bibl. NTIS Order No.: DE82010611. NTIS Prices: PC A03/MF A01.
Report evaluates various methods designed to remove the plenum and the fuel from the TMI-2 reactor during defueling operations. Technical report has immediate use for persons planning TMI-2 decontamination activities.

418. TMI-2 ACCIDENT RECOVERY ACTIVITIES
Examination Results of the TMI Radiation Detector HP-R-213. Prepared by G.M. Mueller for U.S. Dept. of Energy. Springfield, Va.: NTIS, U.S. Dept. of Commerce, 1982. 33p. Bibl. NTIS Order No.: DE82022153. NTIS Prices: PC A03/MF A01.
Technical analysis suggests probable causes of failure for an area radiation detector taken from TMI-2 containment building and estimates the gamma radiation which this device received during the TMI-2 accident. Technical report will be of use to those planning TMI-2 decontamination activities.

419. TMI-2 ACCIDENT RECOVERY ACTIVITIES
Examination Results on TMI-2 Charge Convertors YM-AMP-7023, 2nd YM-AMP-7025. Prepared by M.B. Murphy and R.E. Heintzleman for U.S. Dept. of Energy. Springfield, Va.: NTIS, U.S. Dept. of Commerce, 1982. 28p. Bibl. NTIS Order No.: DE83001661. NTIS Prices: PC A03/MF A01.
Technical report describes probable causes of gamma radiation exposure and degradation of the convertors during the TMI-2 accident. Recommendations for improved instruments are given. Manufacturers of these instruments should be particularly interested in recommendation for instrumental changes.

420. TMI-2 ACCIDENT RECOVERY ACTIVITIES
Feasibility of Vitrifying EPICOR II Organic Resins. Prepared by J.L. Buelt for U.S. Dept. of Energy. Springfield, Va.: NTIS U.S. Dept. of Commerce, 1982. 15p. Bibl. NTIS Order No.: DE83007233. NTIS Prices: PC A02/MF A01.
Laboratory findings suggest EPICOR resins can be successfully destroyed, leaving minimal radioactive waste by-products. Persons planning for TMI-2 decontamination will be most interested in this report.

421. TMI-2 ACCIDENT RECOVERY ACTIVITIES
Feasibility Study: TMI-2 Fuel-recovery Program. Prepared by D.L. Evans for U.S. Dept. of Energy. Springfield, Va.: NTIS, U.S. Dept. of Commerce, 1982. 129p. Bibl. NTIS Order No.: DE83004852. NTIS Prices: PC A07/MF A01.
Evaluates feasibility of constructing a TMI-2 core fuel recovery plant at the Idaho National Engineering Laboratory to recover TMI-2 fuel and to process and package other TMI-2 radioactive wastes. Environmental planners, researchers and persons responsible for TMI-2 decontamination planning will find this report most useful.

422. TMI-2 ACCIDENT RECOVERY ACTIVITIES
Field Measurements and Interpretation of TMI-2 Instrumentation: CF-2-LT2. Prepared by J.E. Jones [et al.] for U.S. Dept. of Energy. Springfield, Va.: NTIS, U.S. Dept. of Commerce, 1982. 44p. Bibl.
NTIS Order No.: DE802010492. NTIS Prices: PC A03/MF A01.
Tests are reported for TMI-2 reactor instrument monitor, located on Core Flood Tank 1A. Report has immediate use for persons planning TMI-2 decontamination activities.

423. TMI-2 ACCIDENT RECOVERY ACTIVITIES
Field Measurements and Interpretation of TMI-2 Instumentation: CF-2-LT4. Prepared by J.E. Jones [et al.] for U.S. Dept. of Energy. Springfield, Va.: NTIS, U.S. Dept. of Commerce, 1982. 44p. Bibl.
NTIS Order No.: DE82010035. NTIS Prices: MF A01.
Tests are reported for TMI-2 reactor instrument monitor, located on Core Flood Tank 1B, to determine if the instrument is functional. Report has immediate use for persons responsible for planning TMI-2 decontamination activities.

424. TMI-2 ACCIDENT RECOVERY ACTIVITIES
Field Measurements and Interpretation of TMI-2 Instrumentation: HI-AMP-2. Prepared by J.E. Jones, [et al.] for U.S. Dept. of Energy. Idaho Falls, Id.: EG and G Idaho, Inc., 1982. 74p. Bibl.
NTIS Order No.: DE82010493. NTIS Prices: PC A04/MF A01.
Describes an evaluation of various TMI-2 instruments following the TMI-2 emergency and indicates equipment is not functioning accurately. Technical report has special use for persons planning TMI-2 recovery activities.

425. TMI-2 ACCIDENT RECOVERY ACTIVITIES
Field Measurements and Interpretation of TMI-2 Instrumentation: HP-R-211. Prepared by J.E. Jones [et al.] for U.S. Dept. of Energy. Springfield, Va.: NTIS, U.S. Dept. of Commerce, 1982. 179p. Bibl.
NTIS Order No.: DE82010046. NTIS Prices: MF A01.
Testing of TMI-2 reactor area radiation monitor is reported. Causes of the monitor's failure are suggested and replacement recommended. Report is immediately useful for persons planning TMI-2 recovery activities.

426. TMI-2 ACCIDENT RECOVERY ACTIVITIES
Field Measurements and Interpretation of TMI-2 Instrumentation: HP-R-212. Prepared by J.E. Jones [et al.] for U.S. Dept. of Energy. Springfield, Va.: NTIS, U.S. Dept. of Commerce, 1982. 63p. Bibl.
NTIS Order No.: DE82010679. NTIS Prices: PC A04/MF A01.
Tests for TMI-2 reactor area radiation monitor are reported. Replacement is recommended, as this monitor, along with others, provides needed radiation readings. Report has immediate use for persons planning TMI-2 recovery activities.

427. TMI-2 ACCIDENT RECOVERY ACTIVITIES
Field Measurements and Interpretation of TMI-2 Instrumentation: HP-R-213. Prepared by J.E. Jones [et al.] for U.S. Dept. of Energy. Springfield, Va.: NTIS, U.S. Dept. of Commerce, 1982. 64p. Bibl.
NTIS Order No.: DE82010678. NTIS Prices: MF A01.
Describes probable causes of failure of radiation monitor HP-R-213, located inside the TMI-2 containment building, and recommends its early replacement. Technical report has particular use for persons planning TMI-2 decontamination activities.

428. TMI-2 ACCIDENT RECOVERY ACTIVITIES
Field Measurements and Interpretation of TMI-2 Instrumentation: HP-R-214. Prepared by J.E. Jones [et al.] for U.S. Dept. of Energy. Springfield, Va.: NTIS, U.S. Dept. of Commerce, 1982. 62p. Bibl.
NTIS Order No.: DE82013780. NTIS Prices: PC A04/MF A01.
Describes probable causes of failure of radiation detector HP-R-214 and recommends early replacement of the instrument. Technical report has immediate use for persons planning TMI-2 decontamination activities.

429. TMI-2 ACCIDENT RECOVERY ACTIVITIES
Field Measurements and Interpretation of TMI-2 Instrumentation: IC-10-DPT. Prepared by J.E. Jones [et al.] for U.S. Dept. of Energy. Springfield, Va.: NTIS, U.S. Dept. of Commerce, 1982. 44p. Bibl.
NTIS Order No.: DE82010037. NTIS Prices: MF A01.
Describes testing and pressure measurements taken on TMI-2 reactor flowmeter, IC-10-DPT. More tests are recommended to determine the reliability of this instrument. Technical report has immediate use for persons planning TMI-2 recovery activities.

430. TMI-2 ACCIDENT RECOVERY ACTIVITIES
Field Measurements and Interpretation of TMI-2 Instrumentation: YM-AMP-7023 and YM-AMP-7025. Prepared by J.E. Jones [et al.] for U.S. Dept. of Energy. Springfield, Va.: NTIS, U.S. Dept. of Commerce, 1982. 74p. Bibl.
NTIS Order No.: DE82010493. NTIS Prices: PC A04/MF A01.
Presents technical evaluation of various damaged TMI-2 instruments, with indication of amounts of damage.

431. TMI-2 ACCIDENT RECOVERY ACTIVITIES
In-vessel Inspection Before Head Removal: TMI II, Phase III (Tooling and Systems Design and Verification). Prepared by G.S. Carter [et al.] for U.S. Dept. of Energy. Springfield, Va.: NTIS, U.S. Dept. of Commerce, 1982. 152p. Bibl.
NTIS Order No.: DE82022519. NTIS Prices: PC E06/MF A01.
Describes preliminary work, equipment needed, and tests performed to determine best methods to remove selected lead screws from TMI-2 reactor so remote handling equipment can view the interior of

the damaged reactor. This report is necessary to the planning of further TMI-2 decontamination and accident recovery activities.

432. TMI-2 HYDROGEN BURN
Investigation of Hydrogen-burn Damage in the TMI Unit 2 Reactor Building. Prepared by N.J. Alvares [et al.] for U.S. Dept. of Energy and EG and G Idaho, Inc. Springfield, Va.: NTIS, U.S. Dept. of Commerce, 1982. 67p. Bibl.
NTIS Order No.: DE82018639. NTIS Prices: PC A04/MF A01.
Discusses photographs of the TMI-2 reactor, showing probable extent of hydrogen burn damage which began about ten hours after the TMI-2 accident began on March 28, 1979. More photographs and data samples are needed. Report will be useful to those persons planning TMI-2 accident recovery operation.

433. TMI-2 ACCIDENT RECOVERY ACTIVITIES
Methods for Eluting Radiocesium from Zeolite Ion Exchange Material in a Column in the TMI-2 Reactor Containment Building. Prepared by J.B. Knauer [et al.] for U.S. Dept. of Energy. Springfield, Va.: NTIS, U.S. Dept. of Commerce, 1982. 18p. Bibl.
NTIS Order No.: DE82018181. NTIS Prices: PC A02/MF A01.
Reviews alternate procedures for separating radiocesium from zeolite ion material as part of TMI-2 radioactive waste processing. Report has use for persons planning TMI-2 decontamination activities.

434. TMI-2 ACCIDENT RECOVERY ACTIVITIES
Nuclear-criticality-safety Studies of Interest to TMI-2 Recovery Operations. Prepared by J.T. Thomas for U.S. Dept. of Energy. Springfield, Va.: NTIS, U.S. Dept. of Commerce, 1982. 22p. Bibl.
NTIS Order No.: DE83000423. NTIS Prices: PC A02/MF A01.
Presents various possible dangerous situations which could develop in course of TMI-2 decontamination activities and suggests nuclear safety precautions to prevent potential dangers. Technical report has immediate interest for persons planning TMI-2 decontamination activities.

435. TMI-2 ACCIDENT RECOVERY ACTIVITIES
Pre-decontamination Gamma-ray Surface Scans in TMI-2 Containment Building, 305' Elevation. Prepared by E.D. Barefoot [et al.] for U.S. Dept. of Energy ... [et al.]. Springfield, Va.: NTIS, U.S. Dept. of Commerce, 1982. 18p. Bibl.
NTIS Order No.: DE82008184. NTIS Prices: MF A01.
Indicates gamma ray scans, performed on December 16, 1982, show a reduction in radiation levels in the TMI-2 containment building. Report has interest for persons planning TMI-2 decontamination activities.

436. TMI-2 ACCIDENT RECOVERY ACTIVITIES
Process Improvement Studies for the Submerged Demineralizer System (SDS) at the TMI Nuclear Power Station Unit 2. Prepared

by D.O. Campbell [et al.] for U.S. Dept. of Energy. Springfield, Va.: NTIS, U.S. Dept. of Commerce, 1982. 30p. Bibl.
NTIS Order No.: DE82014568. NTIS Prices: PC A03/MF A01.
Discusses testing program to evaluate alternative procedures which the Submerged Demineralizer System (SDS) can utilize in processing and decontaminating TMI-2's containment water. Report has immediate use for persons planning TMI-2 decontamination activities.

437. TMI-2 ACCIDENT RECOVERY ACTIVITIES
Reactor-building-basement Radionuclide-distribution Studies.
Prepared by T.E. Cox [et al.] for U.S. Dept. of Energy. Springfield, Va.: NTIS, U.S. Dept. of Commerce, 1982. 25p. Bibl.
NTIS Order No.: DE83002287. NTIS Prices: PC A02/MF A01.
Technical report presents analysis of eight sump samples, taken from TMI-2 containment building in 1981. Findings give radionuclide concentration, based on two separate analyses. Report has direct use for persons planning TMI-2 decontamination and recovery activities.

438. TMI-2 ACCIDENT RECOVERY ACTIVITIES
Recommendations for TMI-2 Instrumentation Surveillance Program. Prepared by J.E. Jones and M.V. Mathis for U.S. Dept. of Energy. Springfield, Va.: NTIS, U.S. Dept. of Commerce, 1982. 88p. Bibl.
NTIS Order No.: DE82012878. NTIS Prices: PC A05/MF A01.
Describes programs of Technical Integration Office (TIO) to examine selected TMI-2 instruments during decontamination process, together with lists of recommended testing equipment. Technical report has direct use for persons planning TMI-2 accident recovery activities.

439. TMI-2 ACCIDENT RECOVERY ACTIVITIES
Review of TMI-2 Resistance Temperature Detectors Accident Data and In-situ Testing. Prepared by J.W. Mock for U.S. Dept. of Energy. Springfield, Va.: NTIS, U.S. Dept. of Commerce, 1982. 29p. Bibl.
NTIS Order No.: DE83004265. NTIS Prices: PC A03/MF A01.
Analyzes data collected from sixteen resistance temperature detectors during and following the TMI-2 accident. Report will be useful for persons planning TMI-2 recovery activities as temperature measurements assist in an understanding of the TMI-2 accident's causes and its development.

440. TMI-2 ACCIDENT RECOVERY ACTIVITIES
A Risk Assessment for the Transportation of Radioactive Ziolite Liners. Prepared by Raymond H.V. Gallucci for U.S. Dept. of Energy. Springfield, Va.: U.S. Dept. of Commerce, 1982. 155p. Bibl.
NTIS Order No.: DE82008034. NTIS Prices: PC A08/MF A01.
OCLC: 9349525

Presents feasibility of shipping TMI-2 radioactive zeolite liners from Pennsylvania to Idaho. Report suggests liners can be safely transported from TMI to Idaho National Engineering Laboratory.

441. TMI-2 ACCIDENT RECOVERY ACTIVITIES
Solid State Track Recorder Neutron Dosimetry Measurements for Fuel Debris Assessment of TMI-2 Demineralizer-A. Prepared by F.H. Ruddy [et al.] for U.S. Dept. of Energy. Springfield, Va.: NTIS, U.S. Dept. of Commerce, 1982. 45p. Bibl.
NTIS Order No.: DE83017004. NTIS Prices: PC A03/MF A01.
Neutron dosimetry measurements evaluate the neutron activity of the TMI-2 fuel in order to estimate the amount of fuel debris remaining in the fuel. Technical report has use for persons planning TMI-2 decontamination and accident recovery activities.

442. TMI-2 ACCIDENT RECOVERY ACTIVITIES
Static in Situ Test of the Axial Power Shaping Rod and Shim Safety Control Rod Mechanisms. Prepared by F.T. Soberano [et al.] for U.S. Dept. of Energy. Springfield, Va.: NTIS, U.S. Dept of Commerce, 1982. 20p. Bibl.
NTIS Order No.: DE82022186. NTIS Prices: PC A02/MF A01.
Describes various technical tests made on eight Axial Power Shaping Rods (APSRs) and sixty-one Shim Safety Control Rods (SSCRs) removed from the TMI-2 reactor core.

443. TMI-2 ACCIDENT RECOVERY ACTIVITIES
Status of TMI-2 Instruments and Electrical Components. Prepared by H.J. Helbert for U.S. Dept. of Energy. Springfield, Va.: NTIS, U.S. Dept. of Commerce, 1982. 18p. Bibl.
NTIS Order No.: DE82021877. NTIS Prices: PC A02/MF A01.
Summarizes collected information concerning TMI-2 instruments and electrical equipment following the TMI-2 accident. Further tests will provide an update to this report. Persons planning TMI-2 recovery operations need to know status of all TMI-2 equipment.

444. TMI-2 PSYCHOLOGICAL EFFECTS OF ACCIDENT
Stressor Effects in Lab and Life: Correspondence Between the Effects of the Accident at TMI and Stress Responses in the Laboratory. Prepared by Siegfried Streufert [et al.] for U.S. Office of Naval Research. Springfield, Va.: NTIS, U.S. Dept. of Commerce, 1982. 22p. Bibl.
NTIS Order No.: Ad-A120 940/2. NTIS Prices: PC A02/MF A01.
Research report concerned with psychological effects of stress upon persons living in vicinity of TMI-2 during the 1979 accident and persons experiencing stress in a laboratory situation. Health professionals will find this report informative.

445. TMI-2 ACCIDENT RECOVERY ACTIVITIES
Task Plan for the U.S. Department of Energy TMI-2 Programs. Prepared by Idaho National Engineering Laboratory for U.S. Dept. of Energy. Springfield, Va.: NTIS, U.S. Dept. of Commerce, 1982. 36p. Bibl.

Reports 147

NTIS Order No.: DE83002815. NTIS Prices: PC A03/MF A01.
Describes U.S. Dept. of Energy programs at TMI-2, focusing on DOE's major responsibilities in TMI-2 clean-up activities: data acquisition, waste immobilization and reactor evaluation. Researchers and persons planning TMI-2 recovery activities will find this summary report particularly useful.

446. TMI-2 ACCIDENT RECOVERY ACTIVITIES
Testing and Examination of TMI-2 Electrical Components and Discrete Devices. Prepared by F.T. Soberano for U.S. Dept. of Energy. Springfield, Va.: NTIS, U.S. Dept. of Commerce, 1982. 82p. Bibl.
NTIS Order No.: DE83000810. NTIS Prices: PC A05/MF A01.
Summarizes findings of tests of various pieces of TMI-2 electrical equipment and presents damage estimates. Report has particular use for persons planning TMI-2 decontamination activities.

447. TMI-2 ACCIDENT RECOVERY ACTIVITIES
TMI-2 Core Examination Program: INEL Facilities-Readiness Study. Prepared by T.B. McLaughlin for U.S. Dept. of Energy. Springfield, Va.: NTIS, U.S. Dept. of Commerce, 1982. 91p. Bibl.
NTIS Order No.: DE83000679. NTIS Prices: PC A04/MF A01.
Views program for examining and storage of TMI-2 radioactive waste materials. Preliminary costs and time schedules for various activities are presented. Report is useful for researchers and planners of TMI-2 decontamination and recovery activities.

448. TMI-2 ACCIDENT RECOVERY ACTIVITIES
Vitrification Process for the Volume Reduction and Stabilization of Organic Resins. Prepared by J.L. Buelt for U.S. Dept. of Energy. Springfield, Va.: NTIS, U.S. Dept. of Commerce, 1982. 28p. Bibl.
NTIS Order No.: DE83002730. NTIS Prices: PC A03/MF A01.
Describes trial tests to determine feasibility of burning and vitrifying ion exchange resins in a single process, as part of radioactive waste processing for water in TMI-2 containment building. This report has particular use for persons planning TMI-2 decontamination activities.

449. TMI-2 ACCIDENT RECOVERY ACTIVITIES
Zeolite Vitrification Demonstration Program. Characterization of Nonradioactive Demonstration Product. Prepared by J.L. Daniel for U.S. Dept. of Energy. Springfield, Va.: NTIS, U.S. Dept. of Commerce, 1982. 43p. Bibl.
NTIS Order No.: DE82022513. NTIS Prices: PC A03/MF A01.
As a part of the TMI-2 Zeolite Vitrification Program for TMI-2 radioactive waste processing, technical report compares ZVDP-4 glass with MCC 76-68 glass for effects of leaching, tensile strength and vaporization at high-temperatures. Report is directed to persons planning TMI-2 decontamination activities.

450. TMI-2 ACCIDENT RECOVERY ACTIVITIES
Zeolite Vitrification Demonstration Program. Nonradioactive Process Operations Summary. Prepared by G.H. Bryan [et al.] for U.S. Dept. of Energy. Springfield, Va.: NTIS, U.S. Dept. of Commerce, 1982. 44p. Bibl.
NTIS Order No.: DE82021310. NTIS Prices: A03/MF A01.
Describes Pacific Northwest Laboratory's Zeolite Vitrification Demonstration Program which developed a glass formulation procedure and demonstrated the vitrification process, using nonradioactive materials. Report is directed to persons planning TMI-2 decontamination activities.

451. TMI-2 ACCIDENT REPORT
Accident at Three Mile Island and its Aftermath. Prepared by A.P. Milinauskas for U.S. Dept. of Energy. Springfield, Va.: NTIS, U.S. Dept. of Commerce, 1982. 35p. Bibl.
NTIS Order No.: DE83012066. NTIS Prices: PC A03/MF A01.
Overhead projections are presented to describe the TMI-2 reactor together with the TMI-2 accident, its consequences and estimated clean-up activities and projected costs. Researchers and persons involved in planning for nuclear power plant safety will find this report useful.

452. TMI-2 RADIATION RELEASES
Aerial Radiological Survey of the TMI Nuclear Station and Surrounding Area, Middletown, Pennsylvania. Prepared by D.P. Colton for U.S. Dept. of Energy. Springfield, Va.: NTIS, U.S. Dept. of Commerce, 1983. 22p. Bibl.
NTIS Order No.: DE84005015. NTIS Prices: PC A02/MF A01.
Results of 1982 aerial radiological survey indicated TMI-2 radioactive wastes were confined to the TMI-2 facility. No evidence of radiation contamination was found in the surrounding area. Ground-based measurements and soil sampling supported aerial survey. Report has immediate use for researchers, environmental planners and persons planning TMI-2 recovery activities.

453. TMI-2 HYDROGEN BURN
Analysis of Air-temperature Measurements from the TMI Unit 2 Reactor Building. Prepared by M.O. Fryer for U.S. Dept. of Energy. Springfield, Va.: NTIS, U.S. Dept. of Commerce, 1983. 41p. Bibl.
NTIS Order No.: DE83011491. NTIS Prices: PC A03/MF A01.
Analyzes TMI-2 ambient air resistance temperature detectors (RTDs) immediately following the hydrogen burn in the TMI-2 reactor. Reactor instruments appear to have performed accurately, with one exception. Report has direct interest for engineers and others planning for safety in nuclear power plants and those persons planning for TMI-2 clean-up.

454. TMI-2 HYDROGEN BURN
Analysis of the TMI Unit 2 Hydrogen Burn. Prepared by J.O.

Henrie and A.K. Postma for U.S. Dept. of Energy. Springfield, Va.: NTIS, U.S. Dept. of Commerce, 1983. 55p. Bibl. NTIS Order No.: DE83012186. NTIS Prices: PC A04/MF A01.

Study analyzes temperature changes, atmospheric pressure changes and the times of these changes in the TMI-2 containment building on March 28, 1979. Photographs and videotapes provide supporting evidence for understanding nature of the hydrogen burn. Researchers dealing with problems of nuclear safety will find this technical analysis useful.

455. TMI-2 ACCIDENT REPORTS
Analysis of TMI Unit 2 Reactor Cooling System Transients.
Prepared by J.O. Henrie and A.K. Postma for U.S. Dept. of Energy. Springfield, Va.: NTIS, U.S. Dept. of Commerce, 1983. 35p. Bibl. NTIS Order No.: DE83017034. NTIS Prices: PC A03/MF A01.

Analyzes probable causes of abrupt transients in TMI-2 reactor cooling system during first day of TMI-2's 1979 accident. Planners of TMI-2 recovery operations will have immediate interest in this report.

456. TMI-2, BIBLIOGRAPHY
Annotated Bibliography of GEND-Sponsored TMI-2 Reports.
Prepared by EG and G Idaho, Inc. for U.S. Dept. of Energy. Springfield, Va.: NTIS, U.S. Dept. of Commerce, 1983. 12p. Bibl.
NTIS Order No.: DE83011438. NTIS Prices: PC A04/MF A01.

Contains brief descriptions of some seventy government-sponsored projects related to the TMI-2 recovery activities.

457. NUCLEAR POWER PLANT SAFETY
Assessment of Mobile Accident Response Capability. Prepared by Waste Management Group, Inc. for U.S. Dept. of Energy. Croton-on-Hudson, N.Y.: The Corporation, 1983. 114p. Bibl.
NTIS Order No.: DE84010860/XAB. NTIS Prices: PC A06/MF A01.
OCLC: 11747344

Report summarizes findings of Mobile Accident Response Capability (MARC) program designed to facilitate response for various types of emergencies at U.S. commercial nuclear power facilities, ranging from the less serious accidents to the most serious TMI-2 crisis. Resources are identified and recommendations are made for design and construction of additional U.S. mobile accident response equipment.

458. TMI-2 ACCIDENT RECOVERY ACTIVITIES
Assessment of the TMI Axial Power-Shaping-Rod Dynamic Test Results. Prepared by R.W. Garner [et al.] for U.S. Dept. of Energy. Springfield, Va.: NTIS, U.S. Dept. of Commerce, 1983. 46p. Bibl. NTIS Order No.: DE83011489. NTIS Prices: PC E04/MF A01.

Describes method designed to insert eight axial power shaping rods in the damaged TMI-2 reactor into a fully inserted position, and compares relevant insertion information with other information on condition of TMI-2 reactor following the 1979 accident. Report has particular use for persons planning TMI-2 recovery operations.

459. TMI-2 ACCIDENT RECOVERY ACTIVITIES
Characterization of EPICOR II Prefilter Liner 3. Prepared by N.L. Wynhoff and V. Pasupathi for U.S. Dept. of Energy. Springfield, Va.: NTIS, U.S. Dept. of Commerce, 1983. 65p. Bibl. NTIS Order No.: DE83012616. NTIS Prices: PC A04/MF A01.
Presents proposed methods of handling the EPICOR II liners, including sampling and analysis techniques and general inspections of the liners for durability. Report will have immediate use for persons in charge of planning for handling these liners, containing TMI-2 radioactive wastes.

460. TMI-2 HYDROGEN BURN
Data Integrity Review of TMI Unit 2. Hydrogen Burn Data. Prepared by J.K. Jacoby [et al.] for U.S. Dept. of Energy. Springfield, Va.: NTIS, U.S. Dept. of Commerce, 1983. 69p. Bibl. NTIS Order No.: DE84004115. NTIS Prices: PC A04/MF A01.
Summarizes statistical data on progression of hydrogen burn inside TMI-2 reactor building during TMI-2 accident. Data were collected from various instruments, and measurements were taken of the reactor building's hydrogen, oxygen, and nitrogen changes. Report has particular use to researchers studying development of the hydrogen burn during the TMI-2 accident.

461. TMI-2 ACCIDENT RECOVERY ACTIVITIES
Design Analysis Report: High Integrity Container for Disposal EPICOR II Prefilter Liners. Prepared by R.L. Chapman and H.W. Reno for U.S. Dept. of Energy. Idaho Falls, Id.: EG and G Idaho, Inc., 1983. 285p. Bibl.
NTIS Order No.: DE83014635. NTIS Prices: PC A13/MF A01.
Presents design for container to be used for permanent storage of liners which contain TMI-2 radioactive wastes and which can be buried in either wet or dry subsurface conditions without leakage. Report has immediate interest for persons planning TMI-2 decontamination activities.

462. TMI-2 RADIATION RELEASES
Estimated Source Terms for Radionuclides and Suspended Particulates During the TMI-2 Defueling Operations Phase II. Prepared by P.G. Voilleque [et al.] for U.S. Dept. of Energy ... [et al.]. Springfield, Va.: NTIS, U.S. Dept. of Commerce, 1983. 60p. Bibl. NTIS Order No.: DE83011877. NTIS Prices: PC A04/MF A01.
As part of the planning for the safe removal of the plenum and fuel from the damaged TMI-2 reactor, this technical report reviews laboratory studies designed to estimate amounts of radiation which must be dealt with during the TMI-2 defueling operations. Possible control techniques are recommended.

463. TMI-2 ACCIDENT RECOVERY ACTIVITIES
EPICOR-II Resin Characterization and Proposed Methods for

Degradation Analysis. Prepared by J.D. Doyle [et al.] for U.S. Dept. of Energy. Springfield, Va.: NTIS, U.S. Dept. of Commerce, 1983. 25p. Bibl.
NTIS Order No.: DE84012170. NTIS Prices: PC A02/MF A01.
Presents preliminary information regarding chemical composition of EPICOR-II resins from TMI-2 reactor and includes proposed plan for their examination. Report has direct use for persons planning TMI-2 decontamination activities.

464. TMI-2 ACCIDENT RECOVERY ACTIVITIES
EPICOR-II Resin/Linear Research Plan. Prepared by J.W. McConnell for U.S. Dept. of Energy. Springfield, Va.: NTIS, U.S. Dept. of Commerce, 1983. 34p. Bibl.
NTIS Order No.: DE83009196. NTIS Prices: PC A03/MF A01.
The Idaho National Engineering Laboratory has the responsibility of examining EPICOR-II prefilters from damaged TMI-2 reactor, and undertaking technical analyses of filters, radioactive waste processing and sample testing. This three-year project has particular interest for researchers and persons responsible for planning for the processing of TMI-2 radioactive wastes.

465. TMI-2 RADIATION RELEASES
Estimated Amount of exp 85 Kr Available for Release from Intact Fuel Rods in the TMI Unit 2 Nuclear Power Station. Prepared by R.A. Lorenz for U.S. Dept. of Energy. Springfield, Va.: NTIS, U.S. Dept. of Commerce, 1983. 28p. Bibl.
NTIS Order No.: DE84001140. NTIS Prices: PC A03/MF A01.
Estimates number of fuel rods surviving the TMI-2 accident without rupturing the amount of Krypton released by the TMI-2 accident, and the probable amount of Krypton still present. Report has immediate use for planners of TMI-2 decontamination activities.

466. TMI-2 ACCIDENT RECOVERY ACTIVITIES
Evaluation of TMI-2 Pressure Switches NM-PS-1454 and NM-PS-4174. Prepared by J.A. Gannon for U.S. Dept. of Energy. Springfield, Va.: NTIS, U.S. Dept. of Commerce, 1983. 45p. Bibl.
NTIS Order No.: DE84004468. NTIS Prices: PC A03/MF A01.
Presents results of examination of alarm switcher, taken from TMI-2 containment building following the TMI-2 emergency. Report recommends specific attention to ways moisture can be prevented from entering the switch enclosure. Report has direct use for equipment designers for U.S. nuclear power facilities.

467. TMI-2 RADIATION STUDIES
Examination Results of the TMI Radiation Detector HP-R-212. Prepared by G.M. Mueller for U.S. Dept. of Energy. Springfield, Va.: NTIS, U.S. Dept. of Commerce, 1983. 45p. Bibl.
NTIS Order No.: DE84004121. NTIS Prices: PC A03/MF A01.
Following examination of area radiation detector HP-R-212 taken from TMI-2 containment building, report suggests probable causes

of failure of the detector. Report has direct use for persons planning TMI-2 decontamination activities.

468. TMI-2 ACCIDENT RECOVERY ACTIVITIES
Final Analysis Data on TMI-2 Reactor Coolant System and Reactor Coolant Bleed-tank Samples. Prepared by T.E. Cox [et al.] for U.S. Dept. of Energy. Springfield, Va.: NTIS, U.S. Dept. of Commerce, 1983. 19p. Bibl.
NTIS Order No.: DE83013538. NTIS Prices: PC A02/MF A01.
Report evaluates and compares TMI-2 reactor coolant samples, reactor coolant bleed-tank samples and steam generator samples made by various organizations at different times. Analytical reports have use in planning TMI-2 decontamination activities.

469. TMI, LAWSUIT REVIEW
General Public Utilities (GPU) Corporation Versus Babcock and Wilcox (B and W) Company Lawsuit Review and Its Effect on Three Mile Island-1 (TMI-1) Nuclear Station Unit 1, Docket 50-289. Vol.1. Washington, D.C.: U.S. Nuclear Regulatory Commission, 1983. 112p. Bibl.
NTIS Order No.: NUREG-1020-VI. NTIS Prices: PC A06/MF A01.
Presents NRC summary of lawsuit record between GPU Corporation and B and W Company to determine if any of the lawsuit record material contradicts the previously endorsed restart of TMI-1. See also entry no. 470.

470. TMI, LAWSUIT REVIEW
General Public Utilities (GPU) Corporation Versus Babcock and Wilcox (B and W) Company Lawsuit Review and Its Effect on Three Mile Island-1 (TMI-1) Nuclear Station Unit 1, Docket 50-289. Vol. 2. Appendices. Washington, D.C.: U.S. Nuclear Regulatory Commission, 1983. 837p. Bibl.
NTIS Order No.: NUREG-1020-V2-APP. NTIS Prices: PC A06/MF A01.
Appendices contain transcript of lawsuit record. See also Volume 1, entry no. 469, for report summary.

471. TMI-2 ACCIDENT RECOVERY ACTIVITIES
Gross Decontamination Experiment Report. Prepared by R. Mason [et al.] for U.S. Dept. of Energy. Springfield, Va.: NTIS, U.S. Dept. of Commerce, 1983. 680p. Bibl.
NTIS Order No.: DE83017046. NTIS Prices: PC A99/MF A01.
Presents technical analysis of decontamination experiment which took place in TMI-2 reactor containment building in March 1982. Report had immediate use for planners of various TMI-2 decontamination activities as it provided a means of evaluating the effectiveness of equipment used.

472. TMI-2 ACCIDENT RECOVERY ACTIVITIES
Mechanistic Debris-Bed Packing Model (PWR). Prepared by J.A. Moore for U.S. Dept. of Energy. Springfield, Va.: NTIS, U.S. Dept. of Commerce, 1983. 21p. Bibl.

Describes fuel debris model for area above and around TMI-2 fuel rod fragments, to provide additional information concerning TMI-2 core meltdown and core condition.

473. TMI-2 ACCIDENT RECOVERY ACTIVITIES
Interim Report on the TMI-2 Purification Filter Examination. Prepared by R.E. Mason for U.S. Dept. of Energy. Springfield, Va.: NTIS, U.S. Dept. of Commerce, 1983. 137p. Bibl.
NTIS Order No.: DE83009199. NTIS Prices: PC A07/MF A01.

Discusses condition of TMI-2 coolant clean-up systems, based on a preliminary examination of the purification filters and the composition of the debris found in these filters. Report is directed to persons planning TMI-2 decontamination and recovery activities.

474. TMI-2 ACCIDENT RECOVERY ACTIVITIES
NDA (Nondestructive Analysis) Measurement of Demineralizers at TMI-2. Prepared by J.R. Phillips [et al.] for U.S. Dept. of Energy. Springfield, Va.: NTIS, U.S. Dept. of Commerce, 1983. 34p. Bibl.
NTIS Order No.: DE84002295. NTIS Prices: PC A03/MF A01.

Describes methods used to estimate fuel debris which accumulated in TMI-2's two demineralizers in the course of the 1979 accident. Report has direct use for persons planning the decontamination and accident recovery activities for TMI-2.

475. TMI-2 ACCIDENT RECOVERY ACTIVITIES
Preparations to Ship EPICOR Liners. Prepared by S.P. Queen for U.S. Dept. of Energy. Springfield, Va.: NITS, U.S. Dept. of Commerce, 1983. 59p. Bibl.
NTIS Order No.: DE83015920. NTIS Prices: PC A04/MF A01.

Discusses gas sampling tool, designed to permit samples to be withdrawn from the EPICOR II liner, and support equipment used for subsequent analysis of TMI-2 radioactive wastes and for determining amounts of radiolytic hydrogen in the liners. Report has immediate use for persons planning TMI-2's decontamination and recovery activities.

476. TMI-2 ACCIDENT RECOVERY ACTIVITIES
Planning Study, Resin and Debris Removal System. TMI Nuclear Station Unit 2 Make-up and Purification Demineralizers. Prepared by E.J. Rankey and W.W. Jenkins for U.S. Dept. of Energy. Springfield, Va.: NTIS, U.S. Dept. of Commerce, 1983. 35p. Bibl.
NTIS Order No.: DE83017009. NTIS Prices: PC A03/MF A01.

Discusses proposed alternative methods of removing radioactive wastes from TMI-2 make-up and purification demineralizers. Report is directed to persons planning TMI-2 decontamination and recovery activities.

477. TMI-2 ACCIDENT RECOVERY ACTIVITIES
Post Accident Decontamination of Reactor Primary Systems and Test Loops. Prepared by C.J. Card for U.S. Dept. of Energy and

Electric Power Research Institute. Palo Alto, Calif.: The Institute, 1983. 108p. Bibl.
NTIS Order No.: DE83007423. NTIS Prices: PC A06/MF A01.
OCLC: 9524468
Describes decontamination procedures taken on four reactor systems contaminated by accident fission products and fuel debris, and relates this information to the decontamination of light water reactors and TMI-2 in particular. Researchers and persons planning for TMI-2 decontamination activities will find this report particularly useful.

478. TMI-2 ACCIDENT RECOVERY ACTIVITIES
Preliminary Radioiodine Source-Term and Inventory Assessment for TMI-2. Prepared by C.A. Pellitier [et al.] for U.S. Dept. of Energy. Springfield, Va.: NTIS, U.S. Dept. of Commerce, 1983. 65p. Bibl.
NTIS Order No.: DE83009735. NTIS Prices: PC A04/MF A01.
OCLC: 12619914
A technical analysis of radioiodine measurements during and following the TMI-2 accident. Data measurements are summarized and mathematical models are developed to account for iodine releases into the atmosphere during the TMI-2 accident. Technical report is directed to those planning for TMI-2 decontamination activities.

479. TMI-2 ACCIDENT RECOVERY ACTIVITIES
Preliminary Report of TMI-2 In-core Instrument Damage. Prepared by M.E. Yancey and N. Wilde for U.S. Dept. of Energy. Springfield, Va.: NTIS, U.S. Dept. of Commerce, 1983. 29p. Bibl.
NTIS Order No.: DE83007281
Suggests TMI-2 reactor core underwent severe damage during the course of the 1979 accident, with the failure of in-core instruments and major changes in the central area of the reactor core. Report is directed to persons planning TMI-2 recovery activities and to researchers.

480. TMI-2 ACCIDENT RECOVERY ACTIVITIES
Proposed Methods of Defueling the TMI-2 Reactor Core. Prepared by J.O. Henrie for U.S. Dept. of Energy. Springfield, Va.: NTIS, U.S. Dept. of Commerce, 1983. 27p. Bibl.
NTIS Order No.: DE83012923. NTIS Prices: PC A03/MF A01.
Report summarizes recommendations from a special group of the U.S. nuclear industry on techniques, equipment, testing, etc. leading to eventual removal of damaged fuel and debris from TMI-2 reactor. Report has direct interest for researchers and the entire U.S. nuclear industry.

481. TMI-2 ACCIDENT RECOVERY ACTIVITIES
Quick-Look Inspection: Results. Prepared by Bechtel Northern Corporation for U.S. Dept. of Energy. Springfield, Va.: NTIS, U.S. Dept. of Commerce, 1983. 63p. Bibl.

Reports 155

NTIS Order No.: DE83012813. NTIS Prices: PC A04/MF A01.
Describes condition of TMI-2 reactor in July and August 1982, based on videotapes of the damaged reactor's interior taken over a three-week period and reviewed by a special group of experts. Report is directed to persons planning TMI-2 recovery activities and to researchers.

482. TMI-2 ACCIDENT RECOVERY ACTIVITIES
Radionuclide Mass Balance for the TMI-2 Accident: Data Base System and Preliminary Mass Balance. Volume 1. Prepared by M.I. Goldman [et al.] for U.S. Dept. of Energy. Springfield, Va.: NTIS, U.S. Dept. of Commerce, 1983. 95p.
NTIS Order No.: DE83012090. NTIS Prices: PC A05/MF A01.
Summarizes research studies by NUS Corporation to develop a computerized data base to support further research of radiological toxic species following TMI-2 accident. Report is directed to researchers and persons planning TMI-2 decontamination activities.

483. TMI-2 ACCIDENT RECOVERY ACTIVITIES
Radionuclide Mass Balance for the TMI-2 Accident. Data Base System and Preliminary Mass Balance. Volume 2. Prepared by M.I. Goldman [et al.] for U.S. Dept. of Energy. Springfield, Va.: NTIS, U.S. Dept. of Commerce, 1983. 169p.
NTIS Order No.: DE83012091. NTIS Prices: PC A08/MF A01.
Presents tables describing radionuclide levels following TMI-2 accident. See also volume 1, entry no 482, for description.

484. TMI-2 ACCIDENT RECOVERY STUDIES
Reactor-Building-Basement Radionuclide and Source Distribution Studies. Prepared by T.E. Cox [et al.] for U.S. Dept. of Energy. Springfield, Va.: NTIS, U.S. Dept. of Commerce, 1983. 27p. Bibl.
NTIS Order No.: DE83014875. NTIS Prices: PC A03/MF A01.
Summarizes data concerning liquid and solid samples taken to support further studies leading to eventual decontamination of TMI-2 Report is directed to persons responsible for TMI-2 decontamination activities and to researchers.

485. TMI-2 ACCIDENT RECOVERY ACTIVITIES
Requirements for Transporting the TMI-2 Core. Prepared by D.E. Watkins for U.S. Dept. of Energy. Springfield, Va.: NTIS, U.S. Dept. of Commerce, 1983. 18p. Bibl.
NTIS Order No.: DE84000859. NTIS Prices: PC A02/MF A01.
Summarizes safety requirements connected with the anticipated transportation of TMI-2 reactor core debris from the accident site to the Idaho National Engineering Laboratory for testing and eventual disposal. Environmental planners and researchers, as well as these persons responsible for TMI-2 decontamination and recovery activities, will have direct interest in this report.

486. TMI-2 ACCIDENT RECOVERY ACTIVITIES
Submerged Demineralizer System Processing of TMI-2 Accident

Waste Water. Prepared by H.F. Sanchez and G.J. Quinn for U.S. Dept. of Energy. Springfield, Va.: NTIS, U.S. Dept. of Commerce, 1983. 65p. Bibl.
NTIS Order No.: DE83009223. NTIS Prices: PC A04/MF A01.
Reviews development of SDS for TMI-2. By August 1982 this system had been responsible for successfully reprocessing 1,000,000 gallons of contaminated water from various locations in TMI-2 containment and auxiliary buildings.

487. TMI-2 ACCIDENT RECOVERY ACTIVITIES
Surface Activity and Radiation Field Measurements of TMI-2 Reactor Building Sgross Decontamination Experiment. Prepared by C.V. McIsaac for U.S. Dept. of Energy [et al.]. Springfield, Va.: NTIS, U.S. Dept. of Commerce, 1983. 122p. Bibl.
NTIS Order No.: DE84003469. NTIS Prices: A06/MF A01.
OCLC: 12619778
Describes sampling procedures, equipment used, and results of analysis of surface samples from concrete and metal surfaces of TMI-2 reactor building in December, 1981 and March, 1982. Researchers and planners for TMI-2 decontamination activities will have interest in this report's technical information.

488. TMI-2 ACCIDENT RECOVERY ACTIVITIES
Technical Assistance and Advisory Group, Report Number 5, Volume 5. Prepared by W.H. Hamilton for U.S. Dept. of Energy. Springfield, Va.: U.S. Dept. of Commerce, 1983. 182p. Bibl.
NTIS Order No.: DE84004467. NTIS Prices: PC A09/MF A01.
Various activities connected with TMI-2 clean-up are discussed, including safeguards to prevent fuel from igniting spontaneously during TMI-2 core head removal; studies to evaluate defueling and core sampling equipment; plenum removal; leadscrew sampling; decontamination procedures for TMI-2 containment building; alternative procedures to transfer TMI-2's damaged fuel canisters.

489. TMI-2 ACCIDENT RECOVERY ACTIVITIES
TMI Cable Tracer Operation and Maintenance Manual for Assembly 417910. Prepared by R.L. Sumstine for U.S. Dept. of Energy. Springfield, Va.: NTIS, U.S. Dept. of Commerce, 1983. 53p.
NTIS Order No.: DE84004945. NTIS Prices: PC A04/MF A01.
Manual gives technical operating instructions for operation and maintenance of a cable tracing system to facilitate locating cables in cable trays for testing.

490. TMI-2 ACCIDENT RECOVERY ACTIVITIES
TMI-2 Core-examination Program: INEL Facilities Readiness Study. Prepared by T.B. McLaughlin for U.S. Dept. of Energy. Springfield, Va.: NTIS, U.S. Dept. of Commerce, 1983. 83p. Bibl.
NTIS Order No.: DE83009200. NTIS Prices: PC A05/MF A01.
Presents updated information concerning remote handling facilities of Idaho National Engineering Laboratory (INEL) designed to

Reports 157

store the TMI-2 reactor core safely and to analyze selected samples from the core debris. Projected costs and projected plan are included. Report is directed to researchers and persons responsible for TMI-2 decontamination activities.

491. TMI-2 ACCIDENT RECOVERY ACTIVITIES
TMI-2 Rission-product Elemental and Isotopic Inventories. Prepared by T.R. England and W.B. Wilson for U.S. Dept. of Energy. Springfield, Va.: NTIS, U.S. Dept. of Commerce, 1983. 5p.
NTIS Order No.: DE83007121. NTIS Prices: PC E03/MF A01.
Presents TMI-2 fission product inventories (total core atoms and kilograms), listed for each nuclide and arranged by element. Calculations are directed to engineers and other responsible for planning TMI-2 decontamination and recovery activities.

492. TMI-2 ACCIDENT RECOVERY ACTIVITIES
TMI-2 Fuel Canister Interface Requirements for INEL. Prepared by D.E. Wilkins for U.S. Dept. of Energy. Springfield, Va.: NTIS, U.S. Dept. of Commerce, 1983. 102p. Bibl.
NTIS Order No.: DE83005746. NTIS Prices: MF A01.
Summarizes specifications for TMI-2 core debris canisters, together with examination and storage requirements. Report is directed to appropriate personnel handling these shipments of TMI-2 core debris and to researchers.

493. TMI-2 ACCIDENT RECOVERY ACTIVITIES
TMI-2 Gadolinia Demonstration Assembly Technical Report. Prepared by L.W. Newmann [et al.] for U.S. Dept. of Energy and Babcock and Wilcox Company. Springfield, Va.: NTIS, U.S. Dept. of Commerce, 1983. 75p. Bibl.
NTIS Order No.: DE83016095. NTIS Prices: A04/MF A01.
OCLC: 14909567
Presents data on condition of TMI-2's reactor core. Core power distributions were monitored using both fixed rhodium and movable gadolinium in-core self-powered neutron detectors. Report is directed to engineers and others planning TMI-2 decontamination activities.

494. TMI-2 ACCIDENT RECOVERY ACTIVITIES
TMI-2 Pressure Transmitter Examination Program. Year-End Report: Examination and Evaluation of Pressure Transmitters CF-1-PT3 and CF-2-LT3. Prepared by R.C. Strahm and M.E. Yancey for U.S. Dept. of Energy. Springfield, Va.: NTIS, U.S. Dept. of Commerce, 1983. 42p. Bibl.
NTIS Order No.: DE83006479. NTIS Prices: PC A03/MF A01.
Describes DOE-sponsored data acquisition program at TMI-2, focusing on detailed laboratory testing of two pressure transmitters removed from TMI-2 reactor building to determine their ability to operate. Technical report is directed to engineers and others responsible for TMI-2 decontamination and recovery activities.

495. TMI-2 RADIATION STUDIES
Use of Multi-Element Beta Dosimeters for Measuring Dose Rates in the TMI-2 Containment Building. Prepared by R.I. Scherpelz [et al.] for U.S. Dept. of Energy. Springfield, Va.: NTIS, U.S. Dept. of Commerce, 1983. 108p. Bibl.
NTIS Order No.: DE84000519. NTIS Prices: PC A06/MF A01.
OCLC: 12606105
Describes and evaluates specially designed multi-element beta dosimeters used to measure beta and gamma radiation releases at numerous locations in the TMI-2 reactor containment building. Three sets of tests were carried out, using over one hundred dosimeters. Report is directed to engineers and others planning TMI-2's decontamination activities.

496. TMI-2 ACCIDENT RECOVERY ACTIVITIES
Assessment of Extent and Degree of Thermal Damage to Polymeric Materials in TMI Unit 2 Reactor Building. Prepared by N.J. Alvares for U.S. Dept. of Energy. Springfield, Va.: NTIS, U.S. Dept. of Commerce, 1984. 59p. Bibl.
NTIS Order No.: DE84010031. NTIS Prices: PC A04/MF A01.
Fire damage within TMI-2 reactor building shows an uneven damage pattern, suggesting burned materials did not experience a uniform exposure to heat during the TMI-2 accident. Report is directed to engineers and others planning for TMI-2's decontamination activities.

497. TMI-2 ACCIDENT RECOVERY ACTIVITIES
Addition of Soluble and Insoluble Neutron Absorbers to the Reactor Coolant System of TMI-2. Prepared by R.F. Hansen [et al.] for U.S. Dept. of Energy [et al.]. Springfield, Va.: NTIS, U.S. Dept. of Commerce, 1984. 52p. Bibl.
NTIS Order No.: DE84014877. NTIS Prices: PC A04/MF A01.
OCLC: 12619844
Indicates the advantage of adding boron compounds to the TMI-2 reactor coolant system, to assist in removal of radioactive wastes from the TMI-2 reactor coolant. Detailed laboratory tests were performed at the Oak Ridge National Laboratory.

498. TMI-2 RADIATION RELEASES
Airborne Cloud Tracking Measurements During the Three Mile Island Nuclear Station Accident, Middletown, Pa. Date of Survey: March-June, 1979. Prepared by R.H. Beers [et al.] for U.S. Dept. of Energy. Middletown, Pa.: EG and G Energy Measurements, 1984. 77p. Bibl.
NTIS Order No.: DE85012060/XAB. NTIS Prices: PC A05/MF A01.
OCLC: 14712246
Presents summary of U.S. Dept. of Energy's aerial radiation monitoring of TMI-2 area during the TMI-2 emergency and measurements of radioactive krypton gas releases during authorized purging operations in June 1980.

499. NUCLEAR POWER PLANT SAFETY
Assessment of Light Water Reactor Safety Since the Three Mile Island Accident. Prepared by Doan L. Phung for U.S. Dept. of Energy. Oak Ridge, Tenn.: Oak Ridge Associated Universities, 1984. 231p. Bibl.
NTIS Order No.: DE85004240/XAB. NTIS Prices: PC A10/MF A01.
OCLC: 12649383
Study suggests NRC and private industry efforts have combined to make U.S. commercial light water nuclear reactors from three to six times safer than reactors were before the TMI-2 emergency in 1979.

500. TMI-2 ACCIDENT RECOVERY ACTIVITIES
Core Activities Program. TMI-2 Core Receipt and Storage Project Plan. Prepared by A.L. Ayers for U.S. Dept. of Energy. Springfield, Va.: NTIS, U.S. Dept. of Commerce, 1984. 45p.
NTIS Order No.: DE85006552/XAB. NTIS Prices: PC E03/MF A01.
Discusses preparations for receiving and storing TMI-2 core debris at Idaho National Engineering Laboratory and the actual transportation and storage operations. Projected costs and schedules for various activities are included.

501. TMI-2 ACCIDENT RECOVERY ACTIVITIES
Design and Operation of the Core Topography Data Acquisition System for TMI-2. Prepared by L.S. Beller and H.L. Brown for U.S. Dept. of Energy. Idaho Falls, Id.: EG and G Idaho, Inc., 1984. 103p. Bibl.
NTIS Order No.: DE84012366. NTIS Prices: PC A06/MF A01.
Presents a description of computer techniques, necessary equipment and investigative procedures used to prepare detailed topographic maps of the TMI-2 reactor cavity following the 1979 accident. Report is directed to engineers and researchers.

502. TMI-2 ACCIDENT RECOVERY ACTIVITIES
Develop a Practical Means to Monitor the Criticality of the TMI-2 Core. Prepared by S.S. Kim [et al.] for U.S. Dept. of Energy. Springfield, Va.: NTIS, U.S. Dept. of Commerce, 1984. 189p. Bibl.
NTIS Order No.: DE85000059/XAB. NTIS Prices: PC AC09/MF A01.
Describes Asymmetric Multiple Position Neutron Source (AMPNS) method and its application to the study of the TMI-2 degraded core through experiments performed on Penn State's Breazeale TRIGA Reactor (PSBR). Findings are directed to engineers and researchers engaged in research on TMI-2 decontamination and TMI-2 accident recovery activities.

503. TMI-2 ACCIDENT RECOVERY ACTIVITIES
Effect of Boron and Gadolinium Concentration on the Calculated Neutron Multiplication Factor of $U(3)O_2$ Fuel Pins in Optimum Geometrics. Prepared by J.T. Thomas; prepared for U.S.

Dept. of Commerce, 1984. Springfield, Va.: NTIS, U.S. Dept. of
Commerce, 1984. 74p. Bibl.
NTIS Order No.: DE85003793/XAB. NTIS Prices: PC A04/MF A01.
Describes the KENO-Va improved Monte Carlo criticality program, used to determine the present chemical composition of TMI-2's anticipated defueling. Report has direct interest for engineers, and researchers and persons planning TMI-2 decontamination activities.

504. TMI-2 ACCIDENT RECOVERY ACTIVITIES
Examination Results of the Three Mile Island Radiation Detector HP-R-212. Prepared by G.M. Mueller for U.S. Dept. of
Energy. Springfield, Va.: NTIS, U.S. Dept. of Commerce, 1984.
42p. Bibl.
NTIS Order No.: DE84007354. NTIS Prices: PC A03/MF A01.
Discusses examination of area radiation detector HP-R-212, formerly located in TMI-2's containment building, and probable reasons for its failure. Report has immediate use for engineers and others planning TMI-2 decontamination activities.

505. TMI-2 ACCIDENT RECOVERY ACTIVITIES
EPICOR-II Resin Characterization and Proposed Methods for
Degradation Analysis. Rev. 1. Prepared by J.D. Doyle [et al.]
for U.S. Dept. of Energy. Springfield, Va.: NTIS, U.S. Dept.
of Commerce, 1984. 25p. Bibl.
NTIS Order No.: DE84014672. NTIS Prices: PC A02/MF A01.
Presents summary of preliminary technical data needed to evaluate resins for degradation in processing of TMI-2 radioactive wastes. Technical report is directed to persons planning TMI-2 decontamination activities.

506. TMI-2 ACCIDENT RECOVERY ACTIVITIES
Equipment for Removal of the TMI-2 Plenum Assembly. Prepared by M.W. Ales [et al.] for U.S. Dept. of Energy. Springfield,
Va.: NTIS, U.S. Dept. of Commerce, 1984. 45p. Bibl.
NTIS Order No.: DE84014442. NTIS Prices: PC A03/MF A01.
Describes special remote handling equipment designed to separate plenum fittings from the TMI-2 reactor core vessel and support shield, to remove fuel debris remnants and to lift and to store the plenum assembly in the shallow part of the TMI-2 refueling canal. Report has significant information for researchers and nuclear safety experts.

507. TMI-2 ACCIDENT RECOVERY ACTIVITIES
Evaluation Results of TMI-2 Solenoids AH-V6 and AH-V74.
Prepared by F.T. Soberano for U.S. Dept. of Energy and EG and
G Idaho, Inc. Philadelphia, Pa.: United Engineers & Constructors,
Inc., 1984. 37p. Bibl.
NTIS Order No.: DE84006085. NTIS Prices: PC A03/MF A01.
Summarizes findings concerning two solenoids in TMI-2 reactor to determine causes of failure. Researchers will have particular interest in recommendations to improve future parts used in nuclear power reactors.

Reports 161

508. TMI-2 ACCIDENT RECOVERY ACTIVITIES
Final Report on the in Situ Testing of Electrical Components and Devices at TMI-2. Prepared by F.T. Soberano for U.S. Dept. of Energy. Philadelphia, Pa.: United Engineers & Constructors, Inc., 1984. 98p. Bibl.
NTIS Order No.: DE8401463. NTIS Prices: PC A06/MF A01.
OCLC: 12619931
Presents summary of performance findings for various pieces of electrical equipment and motors which failed during or following the TMI-2 emergency. Report is directed to engineers and others responsible for TMI-2 recovery activities and to researchers interested in improving nuclear safety equipment.

509. NUCLEAR POWER PLANT SAFETY
Gamma-ray Spectrometer System for High Radiation Fields. Final Report. Prepared by G.R. Lauer [et al.] for U.S. Dept. of Energy and the Electric Power Research Institute. Palo Alto, Calif.: Electric Power Research Institute, 1984. 38p. Bibl.
NTIS Order No.: DE85005800/XAB. NTIS Prices: PC A03/MF A01.
OCLC: 15072080
Describes an improved portable gamma-ray detector system, based on one used at TMI-2, for gamma-ray spectroscopy to measure fission products following serious nuclear reactor accidents.

510. TMI-2 ACCIDENT RECOVERY ACTIVITIES
High Pressure Water Systems for Flushing Loose Fuel Debris from the Reactor Underhead Area at TMI-2. Prepared by H.R. Gardner [et al.] for U.S. Dept. of Energy. Springfield, Va.: NTIS, U.S. Dept. of Commerce, 1984. 84p. Bibl.
NTIS Order No.: DE84015211. NTIS Prices: PC A05/MF A01.
Describes planning and research to develop a remote handling high pressure water system to be used inside TMI-2's reactor pressure vessel to flush off radioactive loose fuel debris from the underhead and plenum cover areas before the TMI-2 reactor removal can be begun.

511. REMOTE CONTROLLED EMERGENCY EQUIPMENT
Improved Robotic Equipment for Radiological Emergencies.
Prepared by C.V. Chester for U.S. Dept. of Energy. Springfield, Va.: NTIS, U.S. Dept. of Commerce, 1984. 57p. Bibl.
NTIS Order No.: DE84017637?XAB. NTIS Prices: PC A04/MF A01.
Since the TMI-2 accident, emergency planning procedures have indicated the need for more efficient remotely controlled equipment to assist with potential nuclear emergency clean-up operations and thereby reduce the danger of radiation exposure for both plant operators and emergency personnel. This report describes recent advances in materials handling equipment and gives approximate development costs.

512. TMI-2 ACCIDENT RECOVERY ACTIVITIES
In-Situ Zeolite Drying. Prepared by G.H. Bryan [et al.] for

U.S. Dept. of Energy. Springfield, Va.: NTIS, U.S. Dept. of Commerce, 1984. 10p. Bibl.
NTIS Order No.: DE84006122. NTIS Prices: PC A02/MF A01.
Describes process whereby SDS liner zeolites can be safely air-dried without generating substantial amounts of hydrogen. After testing, report recommends air drying of the zeolite in the SDS liners prior to shipment.

513. TMI-2 ACCIDENT RECOVERY ACTIVITIES
Involvement of the ORNL Chemical Technology Division in Contaminated Air and Water Handling at the Three Mile Island Nuclear Power Station. Prepared by R.E. Brooksbank and L.J. King for the U.S. Dept. of Energy. Oak Ridge, Tenn.: Oak Ridge National Laboratory, 1984. 61p. Bibl.
NTIS Order No.: ORNL/TM-7044. NTIS Prices: PC A04/MF A01.
Reviews activities of Oak Ridge National Laboratory's Chemical Technology Division in treatment of radioactive wastes and liquid effluents at TMI-2 following the March 1979 emergency. This report was developed at the request of the President's Commission on the Accident at Three Mile Island.

514. TMI-2 ACCIDENT RECOVERY ACTIVITIES
Irradiation Test Report: Foxboro E11GM, Bailey BY3X31A, and Flame Retardant Ethylene Propylene Instrumentation Cable. Prepared by Merlin E. Yancey for U.S. Dept. of Energy. Idaho Falls, Id.: EG and G Idaho, Inc., 1984. 26p. Bibl.
NTIS Order No.: DE85003944?XAB. NTIS Prices: PC A03/MF A01.
Summarizes results of test program designed to determine whether or not pressure transmitters on core flood tanks of TMI-2 reactor functioned throughout the TMI-2 emergency, particularly during periods of highest radiation. Laboratory tests on similar equipment suggest the Bailey Meter Company transmitter was probably more affected by the radiation than the Foxboro units. Associated cabling also showed varying changes due to radiation effects.

515. TMI-2 ACCIDENT RECOVERY ACTIVITIES
Metallurgical Examination of, and Resin Transfer from, Three Mile Island Prefilter Liners. Prepared by J.W. McConnell, Jr. and H.W. Spaletta for U.S. Dept. of Energy [et al.]. Springfield, Va.: NTIS, U.S. Dept. of Commerce, 1984. 72p. Bibl.
NTIS Order No.: DE84016086. NTIS Prices: PC A04/MF A01.
OCLC: 12619927
As part of the planning for decontamination of TMI-2, this technical report summarizes detailed research examination of two EPICOR-II prefilter liners, conducted at the Idaho National Engineering Laboratory, to determine minimum lifetime of the liners containing TMI-2 radioactive waste materials and to make recommendations for testing and storage of future TMI-2 waste materials.

516. TMI-2 ACCIDENT, PSYCHOLOGICAL STRESS
Persistence Differences between the Three Mile Island Residents

and a Control Group. Prepared by Daniel L. Collins for U.S. Air Force. Wright Patterson Air Force Base, Ohio: U.S. Air Force Institute of Technology, 1984. 97p. Bibl.
NTIS Order No.: AD-A145 567/4. NTIS Prices: PC A05/MF A01.
OCLC: 11216748

Research compares psychological responses of a group of TMI area residents with a demographically similar control group from Frederick, Maryland. Research tries to determine if group differences and individual differences within each group could be discerned in persistence in performing routine tasks in a laboratory situation. Testing confirms premise that TMI-2 area residents had a lower rate of persistence with routine tasks. Note: Report was submitted as doctoral dissertation at U.S. Air Force Inst. of Technology.

517. TMI-2 ACCIDENT RECOVERY ACTIVITIES
Plan for Shipment, Storage and Examination of TMI-2 Fuel. Prepared by G.J. Quinn [et al.] for EG and G Idaho, Inc. and U.S. Dept. of Energy. Idaho Falls, Id.: EG and G Idaho, Inc., 1984. 68p. Bibl.
NTIS Order No.: DE84013084. NTIS Prices: PC A04/MF A01.

Summarizes planning for shipment of TMI-2 reactor core debris from TMI site near Middletown, Pa. to the Idaho National Engineering Laboratory, Idaho Falls, Id. Special attention is given to coordinating shipment management details.

518. TMI-2 ACCIDENT RECOVERY ACTIVITIES
Proposed Methods for Defueling the TMI-2 Reactor Core. Prepared by J.O. Henrie for U.S. Dept. of Energy. Richland, Wash.: Rockwell International, Inc., 1984. 27p. Bibl.
NTIS Order No.: DE84010879. NTIS Prices: PC A03/MF A01.

Describes removal techniques and types of equipment for TMI-2 core removal, and also necessary developmental testing recommended by special group of U.S. nuclear industry representatives. This special industry group is known as the Debris Defueling Working Group and was appointed by the U.S. Dept. of Energy.

519. TMI-2 RADIATION RELEASES
Radionuclide Mass Balance for the TMI-2 Accident: Data Through 1979 and Preliminary Assessment of Uncertainties. Prepared by R.J. Davis [et al.] for U.S. Dept. of Energy ... [et al.]. Springfield, Va.: NTIS, U.S. Dept. of Commerce, 1984. 137p. Bibl.
NTIS Order No.: DE85004314/XAB. NTIS Prices: PC E07/MF $5.50.
OCLC: 12619674

Discusses technical development of data base model designed to sample and measure radioisotopes from the TMI-2 accident. Data are included for TMI-2 reactor coolant; make-up, purification, and liquid waste systems; and the TMI-2 reactor building. Preliminary assessment includes information on Strontium, cesium and iodine transfers.

520. TMI-2 ACCIDENT RECOVERY ACTIVITIES
Ramen and Luminescence Spectroscopy of Zirconium Oxide with the Use of the MOLE Microprobe. Prepared by T.E. Doyle and J.L. Alvarez for U.S. Dept. of Energy. Springfield, Va.: NTIS, U.S. Dept. of Commerce, 1984. 13p. Bibl.
NTIS Order No.: DE84017234/XAB. NTIS Prices: PC A02/MF A01.
To assist with TMI-2 decontamination activities, technical report describes use of ramen and luminescence spectroscopy with the MOLE microprobe to identify zirconium oxide's presence and to provide information about temperatures and hydrogen formation in the TMI-2 reactor core.

521. TMI-2 ACCIDENT RECOVERY ACTIVITIES
Results of Analyses Performed on Concrete Cores Removed from Floors and D-Ring Walls of the TMI-2 Reactor Building. Prepared by C.V. McIsaac [et al.] for U.S. Dept. of Energy. Springfield, Va.: NTIS, U.S. Dept. of Commerce, 1984. 44p. Bibl.
NTIS Order No.: DE85001455/XAB. NTIS Prices: PC A03/MF A01.
Analyses of samples taken from surfaces of the interior walls of the TMI-2 reactor building. These September 1983 tests indicate that the epoxy-based nuclear grade paints on the TMI-2 reactor building walls probably prevented significant radionuclide penetration during the TMI-2 accident.

522. NUCLEAR POWER PLANT SAFETY
Review of Light Water Reactor Safety Through the Three Mile Island Accident. Prepared by Doan L. Phung for U.S. Dept. of Energy. Oak Ridge, Tenn.: Oak Ridge Associated Universities, 1984. 140p. Bibl.
NTIS Order No.: DE84012891. NTIS Prices: PC A07/MF A01.
OCLC: 11473135
Suggests nuclear safety at U.S. commercial nuclear power facilities has improved markedly since the 1979 TMI-2 emergency. Problems of U.S. commercial nuclear power facilities in 1979 included: reactor design flaws; insufficient testing of reactors; inconsistent federal regulations; inadequate plant safety procedures and training for plant operators in emergency planning; and inadequate sharing of information on previous nuclear plant emergencies. Study suggests new norms for nuclear plant safety.

523. TMI-2 ACCIDENT RECOVERY ACTIVITIES
Solidification of EPICOR-II Resin Waste Forms. Prepared by R.M. Neilson, Jr. [et al.] for U.S. Dept. of Energy. Springfield, Va.: NTIS, U.S. Dept. of Commerce, 1984. 40p. Bibl.
NTIS Order No.: DE84017410/XAB. NTIS Prices: PC A03/MF A01.
Describes and evaluates alternative methods designed to immobilize ion exchange resin wastes. Report describes planning to solidify TMI-2 EPICOR-II prefilter radioactive wastes using Portland type I-II cement and vinyl ester-styrene.

524. TMI-2 ACCIDENT RECOVERY ACTIVITIES
Submerged Demineralizer System Vessel Shipment Report.

Prepared by G.J. Quinn [et al.] for U.S. Dept. of Energy ... [et al.]. Springfield, Va.: NTIS, U.S. Dept. of Commerce, 1984. 81p. Bibl.
NTIS Order No.: DE84013866. NTIS Prices: PC A05/MF A01.
OCLC: 12619744

As part of the TMI-2 decontamination activities, tests indicate that zeolites and absorbed fission products from the TMI-2 contaminated water were found to generate radiolytic hydrogen and oxygen gases which are potentially flammable. A recombination of these gases with water was found to lower radioactive gas levels and to permit the safe shipment of the radioactive wastes from the TMI-2 site to the Idaho National Engineering Laboratory.

525. TMI-2 ACCIDENT RECOVERY ACTIVITIES
Summary of Radioactive Operations for Zeolite Vitrification Demonstration Program. Prepared by G.H. Bryan [et al.] for U.S. Dept. of Energy. Springfield, Va.: NTIS, U.S. Dept. of Commerce, 1984. 36p. Bibl.
NTIS Order No.: DE84006260. NTIS Prices: PC A03/MF A01.
OCLC: 12619816

A testing program has successfully vitrified the zeolite used in the TMI-2's Submerged Dimineralizer System to a borosilicate glass product. Report summarizes full-scale test of system successfully conducted at Pacific Northwest Laboratory.

526. TMI-2 ACCIDENT REPORTS
Tellurium Release and Deposition During the TMI-2 Accident. Prepared by K. Vinjamuri [et al.] for U.S. Dept. of Energy. Springfield, Va.: NTIS, U.S. Dept. of Commerce, 1984. 56p. Bibl.
NTIS Order No.: DE85006169/XAB. NTIS Prices: PC A04/MF A01.

Discusses probable behavior of tellurium during and following the TMI-2 accident, based on available measurement data and estimated calculations of tellurium releases.

527. TMI-2 ACCIDENT RECOVERY ACTIVITIES
TMI Abnormal Wastes Disposal Options. Prepared by A.L. Ayers, Jr. for U.S. Dept. of Energy. Idaho Falls, Id.: EG and G Idaho, Inc., 1984. 5p. Bibl.
NTIS Order No.: DE84011746. NTIS Prices: PC A02/MF A01.
OCLC: 11721368

Discusses three temporary storage alternatives for the TMI-2 radioactive wastes at the Idaho National Engineering Laboratory. Possible options are: (1) storage in temporary storage casks; (2) storage in underground vaults; or (3) silo storage.

528. TMI-2 ACCIDENT RECOVERY ACTIVITIES
TMI Abnormal Waste Project Plan. Prepared by A.L. Ayers, Jr. for U.S. Dept. of Energy. Springfield, Va.: NTIS, U.S. Dept. of Commerce, 1984. 36p. Bibl.
NTIS Order No.: DE84014715. NTIS Prices: PC A03/MF A01.

Summarizes general planning for disposal of TMI-2 radioactive wastes and outlines interim storage plans, planning procedures, project management recommendations, costs, tentative schedules, etc.

529. TMI-2 ACCIDENT RECOVERY ACTIVITIES
TMI-2 Cable/Connections Program: Fiscal Year 1984 Status Report. Prepared by H.J. Helbert [et al.] for U.S. Dept of Energy. Springfield, Va.: NTIS, U.S. Dept. of Commerce, 1984. 106p. Bibl.
NTIS Order No.: DE85005506/XAB. NTIS Prices: PC A06/MF A01.
Describes tests of some two hundred and thirty-three cable channels within TMI-2's reactor building to assess damage caused by the TMI-2 1979 emergency.

530. TMI-2 ACCIDENT RECOVERY ACTIVITIES
TMI-2 Core Debris-Cesium Release/Settling Test Draft Report. Prepared by D.W. Akers and D.A. Johnson. Springfield, Va.: NTIS, U.S. Dept. of Commerce, 1984. 26p.
NTIS Order No.: DE85003401/XAB. NTIS Prices: PC A03/MF A01.
Testing of samples of TMI-2 core debris were undertaken to estimate probable cesium release, turbidity and potential air pollution.

531. TMI-2 ACCIDENT RECOVERY ACTIVITIES
TMI-2 Core Debris-Cesium Release Settling Test. Prepared by D.W. Akers and D.A. Johnson for U.S. Dept. of Energy. Springfield, Va.: NTIS, U.S. Dept. of Commerce, 1984. 34p. Bibl.
NTIS Order No.: DE5008020/XAB. NTIS Prices: PC A03/MF A01.
Tests on samples of TMI-2 core debris provide information on probable cesium release, turbidity, and radioactive emissions during the TMI-2 reactor defueling.

532. TMI-2 ACCIDENT RECOVERY ACTIVITIES
TMI-2 Core Examination Plan. Prepared by J.O. Carlson for U.S. Dept. of Energy. Springfield, Va.: NTIS, U.S. Dept. of Commerce, 1984. 117p. Bibl.
NTIS Order No.: DE84016303/XAB. NTIS Prices: PC A06/MF A01.
Outlines recommendations, proposed data acquisition methods, and proposed follow-up research relevant to the examination of TMI-2's damaged reactor core.

533. TMI-2 ACCIDENT RECOVERY ACTIVITIES
TMI-2 Fuel Canister Interface Requirements for INEL. Revision 1. Prepared by D.E. Wilkins [et al.] for U.S. Dept. of Energy. Springfield, Va.: NTIS, U.S. Dept. of Commerce, 1984. 90p. Bibl.
NTIS Order No.: DE84014716. NTIS Prices: PC A05/MF A01.
As part of planning for the complete decontamination of TMI-2, this technical report describes canister design requirements for TMI-2 radioactive waste containers for safe storage and handling at the Idaho National Engineering Laboratory.

534. TMI-2 ACCIDENT RECOVERY ACTIVITIES
TMI-2 Core Debris Grab Sample Quick Look Report. Prepared

by D.W. Akers and R.L. Nitachke for U.S. Dept. of Energy. Springfield, Va.: NTIS, U.S. Dept. of Commerce, 1984. 42p. Bibl. NTIS Order No.: DE84011032. NTIS Prices: PC A03/MF A01.

Summarizes examination findings of TMI-2 core debris samples conducted at the Idaho National Engineering Laboratory, to estimate the condition and probable accident damage to the TMI-2 reactor core.

535. TMI-2 ACCIDENT RECOVERY ACTIVITIES
TMI-2 In-core Instrument Damage--an Update. Prepared by M.E. Yancey [et al.] for U.S. Dept. of Energy. Springfield, Va.: NTIS, U.S. Dept. of Commerce, 1984. 78p. Bibl.
NTIS Order No.: DE84015257. NTIS Prices: PC A05/MF A01.

Tests of TMI-2 reactor core instruments, conducted in July and August, 1983, suggest that significant damage to instruments occurred in the central area of the lower reactor vessel while the extension cables for the core instruments appeared in good condition.

536. TMI-2 ACCIDENT RECOVERY ACTIVITIES
TMI-2 (Three Mile Island-2) Leadscrew Debris Pyrophoricity Study. Prepared by R.L. Clark [et al.] for U.S. Dept. of Energy. Springfield, Va.: NTIS, U.S. Dept. of Commerce, 1984. 34p. Bibl.
NTIS Order No.: PC A03/MF A01. NTIS Prices: PC A03/MF A01.

Summary of detailed tests of leadscrew samples from TMI-2 reactor building which evaluate the probability of the core debris igniting spontaneously during the core head removal.

537. TMI-2 ACCIDENT RECOVERY ACTIVITIES
TMI-2 Purification Demineralizer Resin Study. Prepared by J.T. Thompson and T.R. Osterhoudt for GPU Nuclear, Inc. and U.S. Dept. of Energy. Springfield, Va.: NTIS, U.S. Dept. of Commerce, 1984. 30p. Bibl.
NTIS Order No.: DE84014736. NTIS Prices: PC E04/MF $4.75.

Summarizes findings of studies related to TMI-2 make-up and purification system demineralizers, indicating the fuel levels are low enough to permit the demineralizer resins to be removed safely by sluicing through existing plant piping.

538. TMI-2 ACCIDENT RECOVERY ACTIVITIES
TMI-2 Pyrophoricity Studies. Prepared by V. Baston [et al.] for U.S. Dept. of Energy [et al.]. Springfield, Va.: NTIS, U.S. Dept. of Commerce, 1984. 35p. Bibl.
NTIS Order No.: DE85003777/XAB. NTIS Prices: PC E03/MF A01. OCLC: 12619795

Summarizes findings of a relevant literature search and experiments evaluating the necessary safety measures to prevent TMI-2's core debris from igniting spontaneously during the dismantling of the TMI-2 reactor.

539. TMI-2 ACCIDENT RECOVERY ACTIVITIES
TMI-2 Reactor Building Source Term Measurements: Surfaces

and Basement Water and Sediment. Prepared by C.V. McIsaac and
D.G. Keefer for U.S. Dept. of Energy. Springfield, Va.: NTIS,
U.S. Dept. of Commerce, 1984. 90p. Bibl.
NTIS Order No.: DE85003888/XAB. NTIS Prices: PC A05/MF A01.
OCLC: 12619924
   Summarizes findings of radiochemical and element analyses
carried out on samples taken from the external surfaces of the
TMI-2 reactor building, basement water and basement sediment from
August 1979 and December 1983.

540. TMI-2 ACCIDENT RECOVERY ACTIVITIES
   TMI-2 Reactor Vessel Head Removal. Prepared by Paul R.
Bengel [et al.] for U.S. Dept. of Energy [et al.]. Springfield, Va.:
NTIS, U.S. Dept. of Commerce, 1984. 99p. Bibl.
NTIS Order No.: DE85005498/XAB. NTIS Prices: PC A05/MF A01.
OCLC: 12619722
   Describes the removal and storage of the damaged TMI-2 reactor
vessel head. Planning began in July 1982 and successful head removal took place in July 1984 so that the plenum and the reactor
core could be removed in turn.

541. TMI-2 ACCIDENT RECOVERY ACTIVITIES
   Uranium Recovery from a Nuclear Fuel. Prepared by R.L.
Miller [et al.] for U.S. Dept. of Energy. Springfield, Va.: NTIS,
U.S. Dept. of Commerce, 1984. 18p. Bibl.
NTIS Order No.: DE84011363. NTIS Prices: PC A02/MF A01.
   As part of the preparation for potential recovery of uranium
from the damaged TMI-2 reactor, this technical report describes
laboratory tests of two samples of crushed iron-enriched basalt by
three different methods. Test results indicate iron-enriched basalt
can be successfully used to assist with the recovery of uranium
from the TMI-2 fuel.

542. TMI-2 ACCIDENT RECOVERY ACTIVITIES
   Use of Pressurized Water to Decontaminate TMI-2 Leadscrew
Sections. Prepared by H.R. Gardner and L.M. Polentz for U.S. Dept.
of Energy [et al.]. Palo Alto, Calif.: The Institute, 1984. 85p.
Bibl.
NTIS Order No.: DE84015209. NTIS Prices: PC A05/MF A01.
OCLC: 13440666
   Evaluates the capability of water jet flushing operations to
remove fuel debris and radioactive cesium from selected sections of
leadscrew removed from the TMI-2 reactor.

543. TMI-2 ACCIDENT RECOVERY ACTIVITIES
   Zeolite Vitrification Demonstration Program: Characterization
of Radioactive Vitrified Zeolite Materials. Prepared by J.O. Barner
[et al.] for U.S. Dept. of Energy. Springfield, Va.: NTIS, U.S.
Dept. of Commerce, 1984. 80p. Bibl.
NTIS Order No.: DE84007353. NTIS Prices: PC A05/MF A01.
   Summarizes laboratory analyses of the leach behavior of three

sample canisters of TMI-2 radioactive vitrified zeolite materials. Experimental procedures and test results are discussed.

544. TMI-2 ACCIDENT RECOVERY ACTIVITIES
Computer Code Calculations of the TMI-2 Accident: Initial and Boundary Conditions. Prepared by S.R. Behling for U.S. Dept. of Energy. Springfield, Va.: NTIS, U.S. Dept. of Commerce, 1985. 49p. Bibl.
NTIS Order No.: DE85012613/XAB. NTIS Prices: PC A03/MF A01.
Provides necessary information to show basis for development of computer codes used to calculate the severity of the TMI-2 accident.

545. TMI-2 ACCIDENT RECOVERY ACTIVITIES
Decontamination Techniques for Mobile Response Equipment Used at Waste Sites (State-of-the Art Survey). Prepared by J.P. Meade and W.D. Ellis for U.S. Environmental Protection Agency. Springfield, Va.: NTIS, U.S. Dept. of Commerce, 1985. 75p. Bibl.
NTIS Order No.: PB85-247021/XAB. NTIS Prices: PC A04/MF A01.
Reviews use of mobile equipment & protection clothing during hypothetical clean-up operations following a serious nuclear power plant accident. Special attention is given to the clean-up operations at TMI-2 and to the types of clothing and equipment used during these operations.

546. TMI-2 ACCIDENT RECOVERY ACTIVITIES
Disposal Demonstration of a High Integrity Container (HIC) Containing an EPICOR-II Prefilter from Three Mile Island. Prepared by J.W. McConnell [et al.] for U.S. Dept. of Energy. Springfield, Va.: NTIS, U.S. Dept. of Commerce, 1985. 106p. Bibl.
NTIS Order No.: DE85008325/XAB
OCLC: 12619763
Summarizes the loading, transporting and disposal of a demonstration container of low-level radioactive waste materials from TMI-2 at a designated commercial site in the state of Washington. Report has particular significance for nuclear engineers, scientists and environmental planners.

547. TMI-2 RADIATION RELEASES
Environmental Radioactivity at the TMI (Three Mile Island), Venting Phase. Prepared by E. Bretthauer [et al.] for U.S. Environmental Protection Agency. Springfield, Va.: NTIS, U.S. Dept. of Commerce, 1985. 15p.
NTIS Order No.: PB85-227502/XAB. NTIS Prices: PC A02/MF A01.
Indicates that air monitoring of controlled Krypton 85 releases from damaged TMI-2 reactor building showed maximum radioactive exposure to be well within limitations prescribed by health standards. Calculations were monitored at various EPA stations, June 28-July 12, 1980.

548. TMI-2 ACCIDENT RECOVERY ACTIVITIES
Equipment for Removal of the TMI-2 Plenum Assembly. Revision 1. Prepared by W.H. Abbott [et al.] for U.S. Dept. of Energy. Springfield, Va.: NTIS, U.S. Dept. of Commerce, 1985. 72p.
NTIS Order No.: DE85014607/XAB. NTIS Prices: PC A04/MF A01.
Describes special equipment for reactor dismantling designed by B and W first to dislodge the fuel assembly and fittings from the plenum assembly and then to lift and transfer the plenum assembly to the flooded end of the defueling canal.

549. TMI-2 ACCIDENT RECOVERY ACTIVITIES
Examination of HB and BB Leadscrews from TMI Unit 2. Prepared by K. Vinjamuri [et al.] for U.S. Dept. of Energy. Springfield, Va.: NTIS, U.S. Dept. of Commerce, 1985. 297p. Bibl.
NTIS Order No.: DE86002327/XAB. NTIS Prices: PC A13/MF A01.
Discusses detailed examinations of samples removed from TMI-2 reactor's control rod drive leadscrews. Examination suggests significant temperature differences existed between the leadscrews closest to the top and bottom of the plenum assembly as well as differences in surface radionuclide concentrations.

550. TMI-2 ACCIDENT RECOVERY ACTIVITIES
Evaluation of Special Safety Issues Associated With Handling the Three Mile Island Unit 2 Core Debris. Prepared by J.O. Henrie and J.N. Appel for U.S. Dept. of Energy. Springfield, Va.: NTIS, U.S. Dept. of Commerce, 1985. 45p. Bibl.
NTIS Order No.: DE85013763/XAB. NTIS Prices: PC A03/MF A01.
Summarizes various test results and safety analyses related to handling TMI-2 core debris, with particular reference to fire hazards, potential hydrogen production, water removal, canister construction and handling, canister "venting" during storage, etc. Recommendations are designed to ensure safe handling and storage of TMI-2 core debris.

551. NUCLEAR POWER PLANT SAFETY
NRC Policy on Future Reactor Designs: Decisions on Severe Accident Issues in Nuclear Power Plant Regulation. Washington, D.C.: The Commission, 1985. 161p. Bibl.
NTIS Order No.: NUREG-1070/XAB. NTIS Prices: PC A07/MF A01.
OCLC: 14116098
Summarizes NRC's policy statement on nuclear reactor design and safety measures for U.S. commercial nuclear power facilities made as a result of TMI-2 accident findings and safety recommendations.

552. TMI-2 ACCIDENT RECOVERY ACTIVITIES
Possible Options for Reducing Occupational Dose from the TMI-2 Basement. (Rept. for March 1979-December 1984). Prepared by L.F. Munson and R. Hardy for U.S. Nuclear Regulatory Commission. Springfield, Va.: NTIS, U.S. Dept. of Commerce, 1985. 125p. Bibl.

Reports   171

NTIS Order No.: NUREG/CR-4399/XAB. NTIS Prices: PC A06/MF A01.
Presents possible plans for decontaminating the highly contaminated water drained from the basement of the TMI-2 containment building as part of the TMI-2 clean-up activities.

553. TMI-2 ACCIDENT RECOVERY ACTIVITIES
Preliminary Report: Examination of HB and BB Leadscrews from Three Mile Island Unit 2 (TMI-2). Prepared by K. Vinjamuri [et al.] for U.S. Dept. of Energy. Springfield, Va.: NTIS, U.S. Dept. of Commerce, 1985. 243p. Bibl.
NTIS Order No.: DE85010919/XAB. NTIS Prices: PC A11/MF A01.
As part of the general examination of the TMI-2 core, a detailed technical analysis is presented of three leadscrews taken from different positions within the TMI-2 reactor core.

554. TMI-2 ACCIDENT RECOVERY ACTIVITIES
Program Plan for Shipment, Receipt, and Storage of the TMI-2 Core. Revision 1. Prepared by G.J. Quinn [et al.] for U.S. Dept. of Energy [et al.]. Springfield, Va.: NTIS, U.S. Dept. of Commerce, 1985. 56p. Bibl.
NTIS Order No.: DE85008108/XAB. NTIS Prices: PC A04/MF A01.
Changes in TMI-2 plan to ship, store, examine, and ultimately safely dispose of radioactive waste materials from TMI-2 core debris.

555. TMI-2 ACCIDENT RECOVERY ACTIVITIES
Progress Report on the TMI-2 Equipment Examination Program: Beta Radiation Damage Assessment. Prepared by P.R. Bennett for U.S. Dept. of Energy. Springfield, Va.: NTIS, U.S. Dept. of Commerce, 1985. 22p. Bibl.
NTIS Order No.: DE85009991/XAB. NTIS Prices: PC A02/MF A01.
Tests are designed to simulate TMI-2 accident conditions to assess probable damage of TMI-2's electrical equipment from beta radiation exposure.

556. TMI-2 ACCIDENT RECOVERY ACTIVITIES
Solid-State Track Recorder Neutron Dosimetry in the Three Mile Island Unit-2 Reactor Cavity. Prepared by R. Gold [et al.] for U.S. Dept. of Energy. Springfield, Va.: NTIS, U.S. Dept. of Commerce, 1985. 29p. Bibl.
NTIS Order No.: DE85011360/XAB. NTIS Prices: PC A03/MF A01.
Through use of neutron dosimetry and radiation detectors, data estimates indicate at least two tons of fuel debris are at the bottom of the TMI-2 reactor vessel.

557. TMI-2 ACCIDENT REPORTS
Tellurium Chemistry, Tellurium Release and Deposition During the TMI-2 Accident. Prepared by K. Vinjamuri [et al.] for U.S. Dept. of Energy. Springfield, Va.: NTIS, U.S. Dept. of Commerce, 1985. 70p. Bibl.
NTIS Order No.: DE85017500/XAB. NTIS Prices: PC A04/MF A01.

Detailed description of tellurium behavior during and following the TMI-2 accident, based on comparison of results of laboratory tests in various facilities which compared release fractions and samples from TMI-2. Tests suggest only small amounts of tellurium were released.

558. TMI-2 ACCIDENT RECOVERY ACTIVITIES
TMI-2 Defueling System Design Description. Prepared by D.E. Falk and C.E. Swenson for U.S. Dept. of Energy. Springfield, Va.: NTIS, U.S. Dept. of Commerce, 1985. 92p. Bibl.
NTIS Order No.: DE85014611/XAB. NTIS Prices: PC A05/MF A01.
Summary of TMI-2 Dry Defueling System, designed to allow for safe removal of fuel debris from the damaged TMI-2 reactor.

559. TMI-2 ACCIDENT RECOVERY ACTIVITIES
TMI-2 Reactor Vessel Head Removal. Prepared by Paul R. Bengel [et al.] for U.S. Dept. of Energy [et al.]. Springfield, Va.: NTIS, U.S. Dept. of Commerce, 1985. 118p. Bibl.
NTIS Order No.: DE85017613/XAB. NTIS Prices: PC A06/MF A01.
OCLC: 12619722
Continues description of successful removal and storage of TMI-2 reactor vessel head. Radiation exposure during this dismantling of the TMI-2 reactor was less than anticipated.

PART V

U.S. GOVERNMENT-SPONSORED PUBLICATIONS: CONFERENCE PAPERS*

560. TMI-2 RADIATION RELEASES
Plenary Lecture--A Special Radioelement Problem: ORNL Assistance to TMI in Handling Contaminated Air and Water. Prepared by R.E. Brooksbank for U.S. Dept. of Energy. 44p. Bibl.
Research Location: Oak Ridge, Tenn.: Oak Ridge National Laboratory.
Conference: Gatlinburg, Tenn.: 23rd Conf. on Analytical Chemistry in Energy Technology, 9 October 1979.
Order: NTIS, Order No. CONF-791049-26. Prices: PC A03/MF A01.
Summarizes various research activities at Oak Ridge National Laboratory concerned with the handling of contaminated air and water releases resulting from the TMI-2 emergency.

561. TMI-2 RADIATION RELEASES
Atmospheric Release Advisory Capability (ARAC) Response to the Three Mile Island Accident. Prepared by M.H. Dickerson and P.H. Gudiksen for U.S. Dept. of Energy. 5p. Bibl.
Research Location: Livermore, Calif.: Lawrence Livermore Laboratory.
Conference: San Francisco, Calif.: IEEE Nuclear Science Symposium, 17 October 1979.
Order: NTIS, Order No. UCRL-83489. Prices: PC A02/MF A01.
Describes emergency assistance given during March and April, 1979 at the TMI-2 through the U.S. Dept. of Energy's monitoring of radiation releases and the transmission of these data to appropriate agencies.

562. TMI-2 RADIATION RELEASES
Critique of Source Term and Environmental Measurement at TMI. Prepared by A.P. Hull for U.S. Dept. of Energy. 20p. Bibl.
Research Location: Upton, N.Y.: Brookhaven National Laboratory.
Conference: San Francisco, Calif.: IEEE Nuclear Science Symposium, 17 October 1979.
Order: NTIS, Order No. BNL-26970. Prices: PC A02/MF A01.

---

*First line of each entry consists of entry number and subject of the conference paper. Title (underscored) and other bibliographic information follow on subsequent lines.

Describes radiation monitoring of the TMI-2 accident. As the TMI-2 monitors were unable to supply reliable information concerning radiation releases, the existing on-site facilities were supplemented by mobile laboratory facilities of the Pennsylvania Bureau of Radiological Health and the U.S. Department of Energy which tested air, water, soil, vegetation, and milk samples. Some samples were sent other state and federal laboratories for tests.

563. TMI-2 RADIATION RELEASES
Low Cost Radioiodine Air Sampling System for Nuclear Incidents: Experience at TMI. Prepared by C. Distenfield and J. Klemish for U.S. Dept. of Energy. 7p. Bibl.
Research Location: Upton, N.Y.: Brookhaven National Laboratory.
Conference: San Francisco, Calif.: IEEE Nuclear Science Symposium, 17 October 1979.
Order: NTIS, Order No. BNL-27092. Prices: PC A02/MF A01.

Indicates need for inexpensive, rapid air-sampling system to determine presence of iodine isotopes during a future serious nuclear accident. Report pays particular attention to the TMI-2 emergency and the problems of the TMI-2 air-sampling recordings.

564. TMI-2 EMERGENCY PLANNING
Operator/Instrumentation Interactions During the Three Mile Island Accident. Prepared by G.E. Cummings [et al.] for U.S. Dept. of Energy. 4p. Bibl.
Research Location: Livermore, Calif.: Lawrence Livermore Laboratory.
Conference: San Jose, Calif.: IEEE Symposium on Nuclear Power Systems, 19 October 1979.
Order: NTIS, Order No. UCRL-82227. Prices: PC A02/MF A01.

Discusses operator emergency training and instrument problems which contributed to the development of the TMI-2 accident during the first sixteen hours.

565. NUCLEAR POWER PLANT SAFETY
TMI-2 and Its Impact on the Regulatory Process. Prepared by S. Israel for U.S. Nuclear Regulatory Commission.
Research Location: Washington, D.C.: U.S. Nuclear Regulatory Commission.
Conference: Madrid, Spain: CSNI Specialist Meeting on Regulatory Review in the Licensing Process, 7 November 1979.
Order: NTIS, Order No. INIS-MF-5700. Prices: PC A02/MF A01.

Discusses recommendations made to improve emergency planning procedures at U.S. commercial nuclear power plants following publication of various task force findings dealing with the TMI-2 accident.

566. TMI-2 ACCIDENT RECOVERY ACTIVITIES
Demonstration of Alternative Decontamination Techniques at Three Mile Island. Prepared by H.W. Arrowsmith and R.P. Allen for U.S. Dept. of Energy. 41p. Bibl.
Research Location: Richland, Wash.: Battelle Pacific Northwest Laboratories.

Conference Papers 175

Conference: Harrisburg, Pa.: DOE/EPRI Symposium on TMI Reactor Problems, 11 November 1979.
Order: NTIS, Order No. PNL-SA-8143. Prices: PC A03/MF A01.
Discusses various decontamination processes which may be used in processing of TMI-2's radioactive wastes. These methods include: immersion electropolishing, in-situ electropolishing, barrel electropolishing, vibratory finishing, high pressure freon spray, centrifugation, acid absorption, and solidification.

567. TMI-2 ACCIDENT REPORTS
Three Mile Island in Perspective. Prepared by D.B. Trauger for U.S. Dept. of Energy. 12p. Bibl.
Research Location: Oak Ridge, Tenn.: Oak Ridge National Laboratory.
Conference: Paducah, Ky.: Joint Meeting AICHE, ASME, and IEEE, 13 November 1979.
Order: NTIS, Order No. CONF-791131-1. Prices: PC A02/MF A01.
Causes of the TMI-2 accident, the TMI-2 accident development, and planning for TMI-2 decontamination and accident recovery activities are described briefly.

568. TMI-2 ACCIDENT REPORTS
Analysis of Early Core Damage at TMI. Prepared by P.K. Mast [et al.] for U.S. Dept. of Energy. 11p. Bibl.
Research Location: Los Alamos, N.M.: Los Alamos Scientific Laboratory.
Conference: Knoxville, Tenn.: ANS Thermal Reactor Safety Meeting, 8 April 1980.
Order: NTIS, Order No. LA-UR-79-2942. Prices: PC A02/MF A01.
Los Alamos Scientific Laboratory reactor safety groups present detailed analysis of 1979 TMI-2 accident, including estimate of TMI-2 core damage.

569. TMI-2 RADIATION RELEASES
Analysis of the TMI-2 Source Range Detector Response. Prepared by J.F. Carew [et al.] for U.S. Dept. of Energy. 15p. Bibl.
Research Location: Upton, N.Y.: Brookhaven National Laboratory.
Conference: Knoxville, Tenn.: ANS Thermal Reactor Safety Meeting, 8 April 1980.
Order: NTIS, Order No. BNL-NUREG-26762. Prices: MF A01.
Using data supplied by TMI radiation detection equipment, this technical report presents an explanation for the various radical radiation changes which occured in the TMI-2 reactor during the first few hours of the emergency.

570. TMI-2 RADIATION RELEASES
Estimates of Dose to the Population Within Fifty Miles Due to Noble Gas Releases from the TMI Incident. Prepared by C.W. Miller [et al.] for U.S. Dept. of Energy. 8p. Bibl.
Research Location: Oak Ridge, Tenn.: Oak Ridge National Laboratory.

Conference: Knoxville, Tenn.: ANS Thermal Reactor Safety Mting.,
8 April 1980.
Order: NTIS, Order No. CONF-800403-13. Prices: MF A01.
Evaluates the environmental consequences and radiation releases from TMI-2 accident. Conclusions indicate accident caused no long-term health problems for persons living within fifty miles of TMI facility.

571. TMI-2 HEALTH STUDIES
Evaluation of the TMI Accident in the Context of WASH-1400.
Prepared by R.D. Burns, III for U.S. Dept. of Energy. 6p. Bibl.
Research Location: Los Alamos, N.M.: Los Alamos Scientific Laboratory.
Conference: Knoxville, Tenn.: ANS Thermal Reactor Safety Mting.,
8 April 1980.
Order: NTIS, Order No. LA-UR-79-2946. Prices: PC A02/MF A01.
Although the chance of radioactive releases and allied health problems was slight, public opinion perceived the health effects of the TMI-2 accident to be more severe than scientific studies indicated.

572. TMI-2 RADIATION RELEASES
Experiments on Hydrogen for Three Mile Island. Prepared by
R.L. Wooley [et al.] for U.S. Dept. of Energy [et al.]. 11p. Bibl.
Research Location: Utah: Billings Energy Corporation.
Conference: Knoxville, Tenn.: ANS Thermal Reactor Safety Mting.,
8 April 1980.
Order: NTIS, Order No. CONF-800403-7. Prices: PC A02/MF A01.
Summarizes technical tests to determine hydrogen remaining in TMI-2's coolant radioactive water, methods to safely remove the remaining hydrogen and the safest methods of depressurizing the reactor system.

573. TMI-2 ACCIDENT REPORTS
Fuel Damage Estimates for TMI-2 Reactor. Prepared by W.L.
Kirchner [et al.] for U.S. Dept. of Energy. 8p. Bibl.
Research Location: Los Alamos, N.M.: Los Alamos Scientific Laboratory.
Conference: Knoxville, Tenn.: ANS Thermal Reactor Safety Mting.,
8 April 1980.
Order: NTIS, Order No. LA-UR-80-1069. Prices: PC A02/MF A01.
Using TRAC code computer calculations, an estimate is made of the probable damage to the TMI-2 reactor system during the TMI-2 accident.

574. TMI-2 HEALTH STUDIES
Health Effects of the Nuclear Accident at TMI. Prepared by
Jacob I. Fabrikant for U.S. Dept. of Energy [et al.] 24p. Bibl.
Research Location: Berkeley, Calif.: Univ. of California.
Conference: San Francisco, Calif.: Conf. on Environmental Regulation, 18 May 1980.

Order: NTIS, Order No. LBL-11297. Prices: PC A02/MF A01.
Suggests TMI-2 accident produced no longer-term health problems for local area residents. Teenagers, mothers of preschool children, and TMI workers were most depressed by TMI-2 accident. Some TMI plant workers experienced lengthy periods of depression.

575. TMI-2 ACCIDENT REPORTS
OSTG Modeling for the Analysis of the TMI Incident. Prepared by C.J. Hsu. [et al.] for U.S. Dept. of Energy. 5p. Bibl.
Research Location: Upton, N.Y.: Brookhaven National Laboratory.
Conference: Las Vegas, NV.: Ann. Mting of ANS, 8 June 1980.
Order: NTIS, Order No. BNL-NUREG-27185. Prices: PC A02/MF A01.
Describes computer code used for analysis of pressurized water reactor accident for a once-through steam transient generator (OSTG) modeling. Findings regarding TMI-2 accident are summarized, using OSTG model.

576. TMI-2 ACCIDENT RECOVERY ACTIVITIES
Emergency Actions Concerned with Effluent Control at TMI.
Prepared by R.E. Brooksbank, [et al.] for U.S. Dept. of Energy.
Research Location: Ridge, Tenn.: Oak Ridge National Laboratory.
Conference: Los Vegas, N.V.: Ann. Mting of ANS, 8 June 1980.
Order: NTIS, Order No. CONF-800607-60. Prices: PC A03/MF A01.
OCLC: 10238950
Summarizes preparatory planning for safe removal of radioactive water from TMI-2 containment and auxiliary buildings.

577. TMI-2 ACCIDENT REPORTS
System Calculations Related to the Accident at TMI Using TRAC.
Prepared by J.R. Ireland for U.S. Dept. of Energy. 11p. Bibl.
Research Location: Los Alamos, N.M.: Los Alamos Scientific Laboratory.
Conference: Orlando, Fla.: 19th Natl. Heat Transfer Conf., 27 July 1980.
Order: NTIS, Order No. LA-UR-79-3301. Prices: PC A02/MF A01.
Computer calculations, using the Transient Reactor Analysis Code (TRAC-PIA), which enable investigators to simulate the initial development of the TMI-2 accident, are described.

578. TMI-2 ACCIDENT RECOVERY ACTIVITIES
Flowsheet Development Studies for the Decontamination of High-Activity-Level Water at TMI Unit 2. Prepared by E.D. Collins [et al.] for U.S. Dept. of Energy. 33p. Bibl.
Research Location: Oak Ridge, Tenn.: Oak Ridge National Laboratory.
Conference: Portland, Or.: 89th Annual Mting. of Am. Inst. of Chemical Engrs., 17 August 1980.
Order: NTIS, Order No. CONF-800802-13. Prices: PC A03/MF A01.
Summarizes information about several possible chemical processing flowsheets to be used in decontamination of TMI-2 radioactive waste water.

579. NUCLEAR POWER PLANT SAFETY
Recent Chemical Engineering Requirements as a Result of TMI-On-site Experience. Prepared by R.E. Brooksbank for U.S. Dept. of Energy. 24p. Bibl.
Research Location: Oak Ridge, Tenn.: Oak Ridge National Laboratory.
Conference: Portland, Or.: 89th Ann. Mting. of Amer. Insti. of Chemical Engrs., 17 August 1980.
Order: NTIS, Order No. CONF-800802-10. Prices: PC A02/MF A01.

Great technical demands will be placed on chemical engineers to assist with complex TMI-2 recovery activities, in particular radioactive waste processing, and to develop improved reactor safety systems to prevent future nuclear power plant accidents.

580. NUCLEAR POWER PLANT SAFETY
Routine and Post-Accident Sampling in Nuclear Reactors. Prepared by W.J. Armento [et al.] for U.S. Dept. of Energy. 20p. Bibl.
Research Location: Portland, Or.: 89th Ann. Mting. of Am. Insti. of Chemical Engrs. 17 August 1980.
Conference: Portland, Or.: 89 Ann. Mting. of Am. Inst. of Chemical Engrs. 17 August 1980.
Order: NTIS, Order No. CONF-800802-9(OR). Prices: PC A02/MF A01.

Based on TMI-2 accident findings and subsequent safety recommendations for plant emergency warning systems, this report discusses new safety regulations for U.S. commercial nuclear reactor warning systems made following the 1979 TMI-2 emergency.

581. TMI-2 ACCIDENT REPORTS
TMI-2 Criticality Analysis: Analytical Models and Methods. Prepared by J.T. West [et al.] for U.S. Dept. of Energy. 7p. Bibl.
Research Location: Oak Ridge, Tenn.: Oak Ridge National Laboratory.
Conference: Washington, D.C.: NAS International Conf., 17 November 1980.
Order: NTIS, Order No. CONF-801107-29. Prices: PC A02/MF A01.

Summarizes computer calculations and mathematical models used to simulate the disruptive conditions which occurred during the TMI-2 accident.

582. TMI-2 ACCIDENT RECOVERY ACTIVITIES
TMI-2 Criticality Analysis: Parametric Studies and Overall Results. Prepared by R.M. Westfall [et al.] for U.S. Dept. of Energy. 7p. Bibl.
Research Location: Oak Ridge, Tenn.: Oak Ridge National Laboratory.
Conference: Washington, D.C.: ANS International Conf., 17 November 1980.
Order: NTIS, Order No. CONF-801107-32. Prices: PC A02/MF A01.

Conference Papers 179

Summarizes analyses of three disrupted core models to provide necessary preliminary information for persons overseeing TMI-2 accident decontamination and clean-up activities.

583. TMI-1 RESTART
Assessment of Stress-Corrosion Cracking in a Spent-Fuel Pool Pipe. Prepared by R.H. Jones [et al.] for U.S. Dept. of Energy and GPU Service Corporation. 27p. Bibl.
Research Location: Richland, Wash.: Battelle Pacific Northwest Laboratories.
Conference: Toronto, Canada: Corrosion/81 Annual Meeting, 6 April 1981.
Order: NTIS, Order No. DE82010966. Prices: PC A03/MF A01.
Summarizes findings of tests on pipes, with stress corrosion cracks, located in the Spent Fuel Pool Piping System at TMI-1. Report suggests surface defects in the pipe led to the initial cracking.

584. NUCLEAR POWER PLANT SAFETY
Janus Displays for Improved Reactor Plant Operation and Control. By J.J. Feeley for U.S. Dept. of Energy. 7p. Bibl.
Research Location: Idaho Falls, Id.: EG and G Idaho, Inc.
Conference: Miami Beach, Fla.: Ann. Mting. of ANS, 7 June 1981.
Order: NTIS, Order No. DE81026015. Prices: PC A02/MF A01.
Suggests ways whereby computer control systems in large U.S. commercial nuclear power facilities can be improved to give plant operators possible problem situations in a more timely manner. Special reference is made to the TMI-2 emergency problems and the development of the TMI-2 emergency.

585. TMI-2 ACCIDENT REPORTS
Status of the TMI-2 Core. A Review of Damage Assessments.
Prepared by D.W. Croucher for U.S. Dept. of Energy. 15p. Bibl.
Research Location: Idaho Falls, Id.: EG and G Idaho, Inc.
Conference: Sun Valley, Id.: Topical Mting. on Reactor Safety Aspects of Fuel Behavior, 2 August 1981.
Order: NTIS, Order No. DE82005251. Prices: PC A02/MF A01.
Summarizes technical estimates of damage to the TMI-2 reactor during the 1979 emergency. Special attention is paid to hydrogen generation and to the release of fission products during the accident.

586. TMI-2 RADIATION RELEASES
Effects of High Radiation Doses on Linde Ion Siv IE-95. Prepared by N.E. Bibler [et al.] for U.S. Dept. of Energy. 21p. Bibl.
Research Location: Aiken, S.C.: DuPont de Nemours (E.I.) Co., Savannah River Lab.
Conference: Detroit, Mich.: 91st Natl. Mting. of Am. Inst. of Chemical Engrs., 16 August 1981.
Order: NTIS, Order No. DE8108380. Prices: PC A02/MF A01.
Describes investigation findings when the zeolite material, Linde Ion Siv IE-95, is used to remove cesium 137 from contaminated water samples taken from the TMI-2 auxiliary and containment buildings.

587. TMI-2 ACCIDENT RECOVERY ACTIVITIES
Water-Decontamination Process-Improvement Tests and Consideration. Prepared by E.D. Collins [et al.] for U.S. Dept. of Energy. 33p. Bibl.
Research Location: Oak Ridge, Tenn.: Oak Ridge National Laboratory.
Conference: Detroit, Mich.: 91st Natl. Mting. of Am. Inst. of Chemical Engrs., 16 August 1981.
Order: NTIS, Order No. DE81028064. Prices: PC A03/MF A01.
Technical paper suggests that changes in processing of TMI-2 radioactive containment water are necessary to meet interim storage requirements for the radioactive containment water.

588. TMI-2 ACCIDENT RECOVERY ACTIVITIES
TMI Zeolite Vitrification Demonstration Program. Prepared by D.H. Siemens [et al.] for U.S. Dept. of Energy. 12p. Bibl.
Research Location: Richland, Wash.: Battelle Pacific Northwest Laboratories.
Conference: Detroit, Mich.: 91st Natl. Mting. of Am. Inst. of Chemical Engrs., 16 August 1981.
Order: NTIS, Order No. DE81028010. Prices: PC A02/MF A01.
Summarizes program findings relating to development of a glass suitable for immobilizing SDS zeolite, the transfer of zeolite to the in-can melter, and the facilities where the demonstration programs will take place as part of TMI-2 accident recovery activities.

589. TMI-2 ACCIDENT RECOVERY ACTIVITIES
Characterization of Radioactive Ion Exchange Media Waste Generated at Three Mile Island. Prepared by T.C. Runion [et al.] for U.S. Dept. of Energy and Karlsruhe University. 10p. Bibl.
Research Location: Idaho Falls, Id.: EG and G Idaho, Inc.
Conference: Karlsruhe, F.R. Germany: Seminar on Management of Waste from Nuclear Power Plants, 5 October 1981.
Order: NTIS, Order No. DE82007337. Prices: MF A01.
Describes waste processing of contaminated water from the TMI-2 auxiliary and fuel handling building. An Ion exchange system, designated as EPICOR II, was used to process the contaminated water and to separate the cesium and strontium through use of filtration processes.

590. TMI-2 ACCIDENT RECOVERY ACTIVITIES
Development of a High Integrity Container for Storage, Transportation, and Disposal of Radioactive Wastes from TMI Unit II. Prepared by R.E. Holtzworth, [et al.] for U.S. Dept. of Energy. 9p. Bibl.
Research Location: Idaho Falls, Id.: EG and G Idaho, Inc.
Conference: Karlsruhe, F.R. Germany: Seminar on Radioactive Waste from Nuclear Power Plants, 5 October 1981.
Order: NTIS, Order No. DE82005295. Prices: PC A02/MF A01.
Summarizes the DOE-sponsored programs designed to develop a high integrity container to permit TMI-2 radioactive wastes to be disposed of at a commercial site in state of Wasington.

Conference Papers 181

591. TMI-2 RADIATION RELEASES
Venting Krypton-85 from the TMI Unit 2 Reactor Building.
Prepared by H.M. Burton for U.S. Dept. of Energy. 7p. Bibl.
Research Location: Idaho Falls, Id.: EG and G Idaho, Inc.
Conference: Karlsruhe, F.R. Germany: Seminar on Management of Radioactive Waste, 5 October 1981.
Order: NTIS, Order No. DE82007348/XAB. Prices: PC A02/MF A01.
Summarizes details of a longer report titled TMI-2 Reactor Building Purge--KR-85 Venting. The TMI-2 venting program occurred between June 28 and July 11, 1980.

592. TMI-2 ACCIDENT RECOVERY ACTIVITIES
Application of the MOLE in Post-Accident Characterization.
Prepared by S.J. Johnson and J.L. Alvarez for U.S. Dept. of Energy. 12p. Bibl.
Research Location: Idaho Falls, Id.: EG and G Idaho, Inc.
Conference: Gatlinburg, Tenn.: 25th Conf. on Analytical Chemistry and Nuclear Technology, 6 October 1981.
Order: NTIS, Order No. DE82005428. Prices: PC A02/MF A01.
Describes uses of laser spectroscopy in providing chemical information concerning materials remaining in a nuclear reactor following a serious accident. Specific application is made to the TMI-2 emergency.

593. NUCLEAR POWER PLANT SAFETY
Lessons Learned from TMI-2. Prepared by H.J. Zeile for U.S. Dept. of Energy. 7p. Bibl.
Research Location: Idaho Falls, Id.: EG and G Idaho, Inc.
Conference: Augusta, Ga.: DOE Nuclear Facility Conference, 26 October 1981.
Order: NTIS, Order No. DE82005647. Prices: PC A02/MF A01.
Summarizes progress on major safety problems which U.S. commercial nuclear power industry has made since the TMI-2 emergency. Subjects covered include: plant safety, operator training, emergency planning procedures, and ways to improve public information in the event of another serious nuclear power plant accident.

594. TMI-2 ACCIDENT REPORTS
TMI-2 Instrument Analysis Results. Prepared by M.B. Murphy and F.V. Thome for U.S. Dept. of Energy. 12p.
Research Location: Albuquerque, N.M.: Sandia National Laboratories.
Conference: San Francisco, Calif.: ANS Winter Mting., 29 November 1981.
Order: NTIS, Order No. DE81027099. Prices: PC A02/MF A01.
Reviews findings to date of the detailed examination of samples of TMI-2 electrical instruments and equipment following the 1979 TMI-2 emergency. All of the instruments examined had failed even though they were designed to monitor plant conditions during an emergency.

595. TMI-2 ACCIDENT REPORTS
Analysis of the Three Mile Island (TMI-2) Hydrogen Burn.
Prepared by J.O. Henrie and A.K. Postma for U.S. Dept. of Energy.
Research Location: Richland, Wash.: Energy Systems Group.
Conference: Santa Barbara, Calif.: Intl. Topical Mting. on Nuclear
Thermal Hydraulics, 11 January 1982.
Order: NTIS, Order No. DE82020725. Prices: PC A03/MF A01.
Summarizes accident findings concerning the development and continuation of the hydrogen fire during the TMI-2 emergency. Attention is given to recorded temperatures, air pressure and calculations of hydrogen concentrations.

596. TMI-1 RESTART
TRAC Analysis of Steam-Generator Overfill Transients for TMI-1.
Prepared by B. Bassett for U.S. Dept. of Energy. 30p. Bibl.
Research Location: Los Alamos, N.M.: Los Alamos National Laboratory.
Conference: Santa Barbara, Calif.: Intl. Topical Mting. of Nuclear
Thermalhydraulics, 11 January 1982.
Order: NTIS, Order No. DE82017333. Prices: PC A03/MF A01.
Indicates that for the transients investigated, the TMI-1 emergency cooling system has an adequate make-up coolant flow to mitigate an emergency caused by overfilling of the steam generators.

597. TMI-1 RESTART
Evaluation of SCC Test Methods for Inconel 600 in Low Temperature Aqueous Solutions. Prepared by R.C. Newman [et al.] for
U.S. Dept. of Energy. 24p. Bibl.
Research Location: Upton, N.Y.: Brookhaven National Laboratory.
Conference: Washington, D.C.: ASTM Conf. on Environment-
Sensitive Fracture, 1 April 1982.
Order: NTIS, Order No. DE8301358. Prices: PC A02/MF A01.
Discusses testing program designed to locate the causes of stress corrosion cracking of the steam generator tubes at TMI-1. This low temperature stress corrosion cracking began when inner surfaces of the tubes came in contact with the primary coolant. The subsequent testing program took into account various environmental and material problems.

598. TMIk-2 ACCIDENT RECOVERY ACTIVITIES
Applications of Iron-Enriched Basalt to TMI Radioactive Waste
Disposal. Prepared by D.E. Owen [et al.] for U.S. Dept. of Energy.
15p. Bibl.
Research Location: Idaho Falls, Id.: EG and G Idaho, Inc.
Conference: Richland, Wash.: ANS Topical Mting. on Treatment
and Handling of Radioactive Wastes, 19 April 1982.
Order: NTIS, Order No. DE82017764. Prices: PC A02/MF A01.
Indicates the probability that iron-enriched basalt can be used as a medium to immobilize and to dissolve TMI-2 radioactive wastes.

599. TMI-2 ACCIDENT RECOVERY ACTIVITIES
TMI Accident and Post-accident Recovery: What Did We Learn?

Prepared by E.D. Collins for U.S. Dept. of Energy. 33p. Bibl.
Research Location: Oak Ridge, Tenn.: Oak Ridge National Laboratory.
Conference: Oak Ridge, Tenn.: American Society of Cert. Engr. Technicians Mting., 13 May 1982.
Order: NTIS, Order No. DE82015688. Prices: PC A03/MF A01.
Summarizes TMI-2 accident recovery activities and describes progress of TMI-2 decontamination programs to date.

600. TMI-2 ACCIDENT RECOVERY ACTIVITIES
Instrumentation and Electrical Program at the TMI Unit 2, Technical Integration Office. Prepared by L.A. Hecker for U.S. Dept. of Energy [et al.] 12p. Bibl.
Research Location: Idaho Falls, Id.: EG and G Idaho, Inc.
Conference: Dearborn, Mich.: Society of Women Engineers Annual Meeting, 17 June 1982.
Order: NTIS, Order No. DE82018895. Prices: MF A01.
Summarizes various TMI-2 accident recovery activities and gives particular attention to the TMI Instrumentation and Electrical Program. This program identifies and tests various instruments and electrical components to determine reasons for survivability or causes of failure.

601. TMI-2 ACCIDENT RECOVERY ACTIVITIES
Calibration of SSTR Neutron Dosimetry for TMI-2 Applications.
Prepared by R. Gold [et al.] for U.S. Dept. of Energy. 29p. Bibl.
Research Location: Richland, Wash.: Hanford Engineering Development Laboratory.
Conference: Richland, Wash.: Mting. of Pacific NW Working Group on Nuclear Track Registration, 28 July 1982.
Order: NTIS, Order No. DE83004073. Prices: PC A03/MF A01.
Suggests methods whereby neutron dosimetry can be used to follow and to identify fuel debris in the TMI-2 reactor core and in the primary coolant system.

602. TMI-2 ACCIDENT RECOVERY ACTIVITIES
TMI-2 Core Examination Plan. Prepared by D.E. Owen [et al.] 11p. Bibl.
Research Location: Idaho Falls, Id.: EG and G Idaho, Inc.
Conference: Chicago, Ill.: Intl. Mting. on Thermal Nuclear Reactor Safety, 29 August 1982.
Order: NTIS, Order No. DE83000565. Prices: PC A02/MF A01.
Summarizes detailed planning for examination of TMI-2 reactor core: before reactor head removal; before plenum removal; during defueling; and off-site examinations. All examinations have the long-range goal of developing nuclear reactors with improved safety features to prevent future accidents.

603. TMI-2 ACCIDENT RECOVERY ACTIVITIES
Development of High-Integrity Container for Intermediate Depth Burial of Special Waste. Prepared by R.L. Chapman [et al.] for U.S. Dept. of Energy and Nuclear Packaging, Inc. 6p. Bibl.

Research Location: Idaho Falls, Id.: EG and G Idaho, Inc.
Conference: Winnipeg, Canada: Int'l. Conference on Radioactive
 Waste Management, 12 Sept. 1982. 6p. Bibl.
Order: NTIS, Order No. DE83000593. Prices: PC A02/MF A01.
 Describes the characteristics of containers designed to hold
the EPICOR-II liners which were used to decontaminate radioactive
water spilled into the Auxiliary and Fuel Handling Building during
the TMI-2 accident. Design requirements for containers expect to
keep these TMI-2 radioactive wastes isolated and immobile for 300
years or longer.

604. TMI-2 ACCIDENT RECOVERY ACTIVITIES
 EPICOR-II Research and Disposition Program at the INEL.
Prepared by H.W. Reno and R.L. Dodge for U.S. Dept. of Energy.
15p. Bibl.
Research Location: Idaho Falls, Id.: Idaho National Engineering
 Laboratory.
Conference: San Diego, Calif.: ANS Workshop on LLW Packaging
 and Shipping, 12 Sept. 1982.
Order: NTIS, Order No. DE83001352. Prices: PC A02/MF A01.
 Describes the treatment of 50 EPICOR-II liners, used to remove
radioactive wastes from the TMI-2 waste water. Subsequent plans
for treatment of the EPICOR-II liners are presented.

605. TMI-2 ACCIDENT RECOVERY ACTIVITIES
 Leach Behavior and Mechanical-Integrity Studies of Irradiated
EPICOR-II Waste Products. Prepared by R.E. Barletta [et al.] for
U.S. Dept. of Energy. 5p. Bibl.
Research Location: Upton, N.Y.: Brookhaven National Laboratory.
Conference: Winnipeg, Canada: Int'l. Conference on Radioactive
 Waste Management, 12 Sept. 1982.
Order: NTIS, Order No. DE83001308. Prices: MF A01.
 Technical report indicates no harmful fission product radiation
releases will occurr when mixtures, similar to TMI-2 radioactive
waste products, are released into cement composites.

606. TMI-2 ACCIDENT RECOVERY ACTIVITIES
 Initial SCDAP Predictions of the TMI-2 Event. Prepared by
C.M. Allison [et al.] for U.S. NRC and U.S. Dept. of Energy. 9p.
Bibl.
Research Location: Idaho Falls, Id.: EG and G Idaho, Inc.
Conference: Gaithersburg, Md.: 10th Water Reactor Safety Info.
 Conf., 12 October 1982.
Order: NTIS, Order No. DE83003238. Prices: PC A02/MF A01.
 Summarizes development of Severe Damage Analysis Package
(SCDAP) Computer Code at the Idaho National Engineering Laboratory
to simulate core damage of TMI-2 reactor during the TMI-2 emergency.

607. TMI-2 ACCIDENT RECOVERY ACTIVITIES
 Technology Transfer at TMI Unit 2. Prepared by H.M. Burton
and W.W. Bixby for U.S. Dept. of Energy. 6p. Bibl.

Conference Papers   185

Research Location: Idaho Falls, Id.: EG and G Idaho, Inc.
Conference: Buenos Aires, Argentina: Int'l. Conference on Nuclear Technology Transfer, 1 November 1982.
Order: NTIS, Order No. DE83005278. Prices: PC A02/MF A01.

The U.S. Department of Energy's goals for TMI-2 clean-up and recovery activities are summarized, with emphasis on improved commercial nuclear plant safety in light of TMI-2 accident findings, progress in technology related to decontamination work at nuclear facilities and improvements in design and operation of U.S. commercial nuclear power plants.

608. TMI-2 ACCIDENT RECOVERY ACTIVITIES
Analysis of TMI-2 Samples Using the Molecular Optical Laser Examiner. Prepared by T.E. Doyle and J.L. Alvarez for U.S. Dept. of Energy. 11p. Bibl.
Research Location: Idaho Falls, Id.: Eg and G Idaho, Inc.
Conference: Washington, D.C.: ANS Winter Mting., 14 November 1982.
Order: NTIS, Order No. DE83005286. Prices: PC A02/MF A01.

Describes detailed examination of TMI-2 core debris samples to determine chemical changes occurred during the TMI-2 emergency and to understand the TMI-2 accident so that appropriate clean-up activities can be initiated.

609. TMI-2 ACCIDENT RECOVERY ACTIVITIES
Axial-Power-Shaping Rod Insertion Test. Prepared by K. Parlee [et al.] for U.S. Dept. of Energy [et al.]. 5p. Bibl.
Research Location: Idaho Falls, Id.: EG and G Idaho, Inc.
Conference: Washington, D.C.: ANS Winter Mting., 14 November 1982.
Order: NTIS, Order No. DE83005543. Prices: PC A02/MF A01.

Describes performance testing of eight TMI-2 axial-power-shaping rods (APSRs) to assist in estimating damage to TMI-2's upper plenum and reactor core. Test results will assist in planning further TMI-2 recovery activities.

610. TMI-2 ACCIDENT RECOVERY ACTIVITIES
Control of Radiolytic Gases in Liners of Radioactive Zeolites. Prepared by J. Greenborg [et al.] for U.S. Dept. of Energy [et al.]. 5p. Bibl.
Research Location: Parsippany, N.J.: General Public Utilities, Inc.
Conference: Washington, D.C.: ANS Winter Mting., 14 November 1982.
Order: NTIS, Order No. DE83005546. Prices: PC A02/MF A01.

Discusses methods to control production of radiolytic gases in the submerged demineralizer liners, designed for shipping, storage, and ultimate disposal of TMI-2 radioactive wastes.

611. TMI-2 ACCIDENT RECOVERY ACTIVITIES
Dose Reduction and Contamination Control in the TMI-2 Reactor Building. Prepared by G.R. Eidam and D.W. Leigh for Bechtel Intl. Corp. and U.S. Dept. of Energy. 4p. Bibl.

Conference: Washington, D.C.: ANS Winter Mting., 14 November 1982.
Order: NTIS, Order No. DE83005541. Prices: PC A02/MF A01.
Describes 1982 experiment to decontaminate selected areas of the TMI-2 reactor building. Hydrolasing, a pressurized-water spray method, and other decontamination techniques were tested.

612. TMI-2 ACCIDENT RECOVERY ACTIVITIES
Fission Product Transfer in the TMI-2 Purification System. Prepared by T.E. Cox for U.S. Dept. of Energy. 4p. Bibl.
Research Location: Idaho Falls, Id.: Eg and G Idaho, Inc.
Conference: Washington, D.C.: ANS Winter Mting., 14 November 1982.
Order: NTIS, Order No. DE83005547. Prices: PC A02/MF A01.
A summary report describes the TMI-2 purification system which operated during the 1979 TMI-2 accident. As part of the TMI-2 recovery activities, an analysis is made of demineralizer resins and filters to determine their condition and the composition of materials retained by the resins and filters.

613. TMI-2 ACCIDENT RECOVERY ACTIVITIES
Instrumentation and Electrical Program at TMI-2. Prepared by W.F. Schwarz for U.S. Dept. of Energy. 3p. Bibl.
Research Location: Idaho Falls, Id.: EG and G Idaho, Inc.
Conference: Washington, D.C.: ANS Winter Mting., 14 November 1982.
Order: NTIS, Order No. DE83005542. Prices: PC A02/MF A01.
Summarizes condition and operability of some 600 electrical instruments located in the TMI-2 containment building.

614. TMI-2 ACCIDENT RECOVERY ACTIVITIES
Investigation of Hydrogen-Burn Damage in the TMI-2 Reactor Building. Prepared by N.L. Alvares [et al.] for U.S. Dept. of Energy. 5p. Bibl.
Research Location: Livermore, Calif.: Lawrence Livermore National Laboratory.
Conference: Washington, D.C.: ANS Winter Mting., 14 November 1982.
Order: NTIS, Order No. DE83005548. Prices: PC A02/MF A01.
Summarizes findings of fifteen entries into the TMI-2 reactor building together with photographs and video recordings. Preliminary findings indicate a burn pattern and overpressure damage throughout the TMI-2 reactor building.

615. TMI-2 ACCIDENT RECOVERY ACTIVITIES
Overview of the TMI-2 Core-Examination Plan. Prepared by K.C. Sumpter [et al.] for U.S. Dept. of Energy. 3p. Bibl.
Research Location: Idaho Falls, Id.: EG and G Idaho, Inc.
Conference: Washington, D.C.: ANS Winter Mting., 14 November 1982.
Order: NTIS, Order No. DE83005280. Prices: PC A02/MF A01.

Summarizes program details for defueling and analyzing TMI-2 debris, leading to better understanding of the TMI-2 accident and the recovery and clean-up measures needed for accident recovery.

616. TMI-2 RADIATION RELEASES
Preliminary Results of the TMI-2 Radioactive Iodine Mass Balance Study. Prepared by C.A. Pelletier [et al.] for U.S. Dept. of Energy. 6p. Bibl.
Research Location: Idaho Falls, Id.: EG and G Idaho, Inc.
Conference: Washington, D.C.: ANS Winter Mting., 14 November 1982.
Order: NTIS, Order No. DE83005540. Prices: PC A02/MF A01.
Summarizes findings from samples taken from the TMI-2 reactor building after the emergency. Samples indicate iodine released in TMI-2 reactor building is probably less than originally estimated.

617. TMI-2 ACCIDENT RECOVERY ACTIVITIES
Research and Disposition of Highly Loaded Organic Resins.
Prepared by R.C. Schmitt [et al.] for U.S. Dept. of Energy. 18p. Bibl.
Research Location: Idaho Falls, Id.: EG and G Idaho, Inc.
Conference: Washington, D.C.: ANS Winter Mting., 14 November 1982.
Order: NTIS, Order No. DE83005274. Prices: PC A02/MF A01.
Discusses details of TMI-2 debris' radioactive waste processing, emphasizing the problems connected with the transportation of the EPICOR-II liners from TMI to the Idaho National Laboratory. Special attention is given to testing of wastes and to development of containers for TMI-2 waste disposal.

618. TMI-2 ACCIDENT RECOVERY ACTIVITIES
Special Handling Requirements for Highly Loaded Organic Resins.
Prepared by R.E. Ogle [et al.] for GPU Nuclear, Inc. and U.S. Dept. of Energy. 4p. Bibl.
Research Location: Idaho Falls, Id.: EG and G Idaho, Inc.
Conference: Washington, D.C.: ANS Winter Mting., 14 November 1982.
Order: NTIS, Order No. DE83005545. Prices: PC A02/MF A01.
Summarizes findings regarding shipment of a sample prefilter liner, PF-16, containing TMI-2 radioactive waste water, from TMI to Idaho National Laboratory.

619. TMI-2 ACCIDENT RECOVERY ACTIVITIES
Surface Disposition Measurements of the TMI-2 Gross Decontamination Experiment. Prepared by C.V. McIssac and D.C. Hetzer for U.S. Dept. of Energy. 5p. Bibl.
Research Location: Idaho Falls, Id.: EG and G Idaho, Inc.
Conference: Washington, D.C.: ANS Winter Mting., 14 November 1982.
Order: NTIS, Order No. DE83005387. Prices: PC A02/MF A01.
Summary of a water spray decontamination technique carried out

in TMI-2's reactor building. Sampling techniques connected with the experiment are described.

620. TMI-2 ACCIDENT RECOVERY ACTIVITIES
Technology Transfer in the DOE TMI Research Programs.
Prepared by F.L. Meltzer and B.A. Ettinger for U.S. Dept. of Energy. 4p. Bibl.
Research Location: Idaho Falls, Id.: EG and G Idaho, Inc.
Conference: Phoenix, Az.: ASME Winter Annual Mting., 14 November 1982.
Order: NTIS, Order No. DE83005396. Prices: PC A02/MF A01.
Summarizes activities of joint U.S. government and U.S. nuclear power industry to improve nuclear power plant safety and to continue TMI-2 accident recovery activities. Communication methods include a newsletter, technical reports, seminars, papers, journal articles, videotapes, and two computer conferencing networks.

621. TMI-2 ACCIDENT RECOVERY ACTIVITIES
TMI-2 Quick Look Examination. Prepared by W.A. Franz [et al] for U.S. Dept. of Energy [et al.]. 4p. Bibl.
Research Location: Idaho Falls, Id.: EG and G Idaho, Inc.
Conference: Washington, D.C.: ANS Winter Mting., 14 November 1982.
Order: NTIS, Order No. DE83005544. Prices: PC A02/MF A01.
Summarizes planning to gain earliest possible access to the TMI-2 reactor vessel for evaluating the condition of both the plenum assembly and the reactor core. The Quick Look Technique involves removal of a CRDM leadscrew and inserting a closed circuit television camera in that vacant space for remote viewing of the interior of the TMI-2 reactor.

622. TMI-2 ACCIDENT RECOVERY ACTIVITIES
TMI-2 Reactor-Vessel Head Removal and Damaged-Core-Removal Planning. Prepared by J.A. Logan [et al.] for U.S. Dept. of Energy. 4p. Bibl.
Research Location: Idaho Falls, Id.: EG and G Idaho, Inc.
Conference: Washington, D.C.: ANS Winter Mting., 14 November 1982.
Order: NTIS, Order No. DE83005393. Prices: PC A02/MF A01.
Describes anticipated benefits of examination of TMI-2 reactor vessel after dismantling. Data will be available concerning the effects of the damaged coolant on fuel cladding, fuel materials, etc. and for estimating temperature sequences during the TMI-2 accident in various areas of the reactor core.

623. TMI-2 ACCIDENT RECOVERY ACTIVITIES
Immobilization of TMI Core Debris. Prepared by J.M. Welch [et al.] for U.S. Dept. of Energy. 19p. Bibl.
Research Location: Idaho Falls, Id.: EG and G Idaho, Inc.
Conference: Chicago, Ill.: Ann. Mting. of American Ceramic Society, 25 April 1983.

Conference Papers   189

Order: NTIS, Order No. DE83013540. Prices: PC A02/MF A01.
    Discusses the possible solidification of the TMI-2 core debris
in iron-enriched basalt (IEB).

624. NUCLEAR POWER PLANT SAFETY
    Nuclear Reactor Safety in the U.S.A. Prepared by J.C. Vigil
for U.S. Dept. of Energy. 11p. Bibl.
Research Location: Los Alamos, N.V.: Los Alamos National Laboratory.
Conference: Ventura, Calif.: Annual Symposium of Mexican Amer.
    Engr. Society, 3 May 1983.
Order: NTIS, Order No. DE83011182. Prices: PC A02/MF A01.
    Suggests that advanced computer simulations at Los Alamos National
Laboratory have substantially reduced risk of another serious U.S.
commercial nuclear power plant emergency like the 1979 TMI-2 accident. Public perception of nuclear power risks does not take into
account the improved public health precautions and improved safety
standards for nuclear power reactors made since the TMI-2 accident.

625. TMI-2 ACCIDENT RECOVERY ACTIVITIES
    Evolution of Zeolite Mixtures for Decontamination Activity Level
Water in the Submerged Demineralizer System (SDS) Flowsheet at the
TMI Nuclear Power Station, Unit 2. Prepared by L.J. King [et al.]
for U.S. Dept. of Energy. 16p. Bibl.
Research Location: Reno, Nev.: Intl. Zeolite Conference, 10 July
    1983.
Conference: Reno, Nev.: Intl. Zeolite Conference, 10 July 1983.
Order: NTIS, Order No. DE83012203. Prices: PC A02/MF A01.
    Mixtures of Linde Ionsiv IE-96 and Ionsiv A-51 zeolites were
tested for their effective use in the Submerged Demineralizer System
(SDS) to decontaminate approximately 700,000 gallons of radioactive
water in TMI-2's containment building.

626. TMI-2 ACCIDENT RECOVERY ACTIVITIES
    Micro-Ramen Analysis of TMI Samples. Prepared by T.E.
Doyle and J.L. Alvarez for U.S. Dept. of Energy. 7p. Bibl.
Research Location: Idaho Falls, Id.: EG and G Idaho, Inc.
Conference: Phoenix, Az.: 18th Ann. Microbeam Analysis Society
    Conf., 8 August 1983.
Order: NTIS, Order No. DE84001003. Prices: PC A02/MF A01.
    By using a Ramen microprobe, various TMI-2 samples are
analyzed. Particular attention is given to identification of zirconium
oxide and particles in TMI-2's purification system.

627. TMI-1 RESTART
    Mechanism of TMI-1 Steam Generator Failure. Prepared by
R.C. Newman [et al.] for U.S. Dept. of Energy. 22p. Bibl.
Research Location: Upton, N.Y.: Brookhaven National Laboratory.
Conference: Myrtle Beach, S.C.: Environmental Degradation of
    Materials in Nuclear Power Systems Water Reactors, 22 August
    1983.

Order: NTIS, Order No. DE83015770. Prices: PC A02/MF A01.
Discusses relevance of various laboratory tests to determine causes of TMI-1 steam generator cracking.

628. TMI-2 ACCIDENT REPORTS
Evaluation of Past Nuclear Events and CORRAL-2 Analyses of Fission-Product Transport at TMI-2 and PRTR. Prepared by D.E. Kudera [et al.] for U.S. Dept. of Energy. 7p. Bibl.
Research Location: Idaho Falls, Id.: EG and G Idaho, Inc.
Conference: Cambridge, Mass.: Intl. Mting. on LWR Severe Accident Evaluation, 28 August 1983.
Order: NTIS, Order No. DE84000889. Prices: PC A02/MF A01.
Discusses serious nuclear facility accidents and reviews the TMI-2 accident sequence. Using the CORRAL-2 computer code, calculations are presented for probable fission product releases during the TMI-2 emergency.

629. TMI-2 ACCIDENT REPORTS
Fuel Models and Results from the TRAC-PF1/MIAS TMI-2 Accident Calculation. Prepared by E.C. Schwegler and P.J. Maudlin for U.S. Dept. of Energy. 4p. Bibl.
Research Location: Los Alamos, N.M.: Los Alamos National Laboratory.
Conference: Cambridge, Mass.: Int'l. Mting. on Light-Water Reactor Severe Accident Evaluation, 28 August 1983.
Order: NTIS, Order No. DE83005240. Prices: PC A02/MF A01.
Describes various fuel models used to analyze development of TMI-2 accident and suggests causes for fuel rod behavior changes.

630. TMI-2 ACCIDENT RECOVERY ACTIVITIES
TMI-2 (Three Mile Island Unit 2) Core Examination. Prepared by R.R. Hobbins [et al.] for U.S. Dept. of Energy. 9p. Bibl.
Research Location: Idaho Falls, Id.: EG and G Idaho, Inc.
Conference: Cambridge, Mass.: Intl. Mting. on LWR Severe Accident Evaluation, 28 August 1983.
Order: NTIS, Order No. DE84000888. Prices: PC A02/MF A01.
Summarizes details of TMI-2 damaged reactor core examination, with particular attention to safety precautions and to goals and techniques used for each reactor core examination.

631. TMI-2 ACCIDENT REPORTS
TMI-2 (Three Mile Island Unit 2) Core Damage: A Summary of Present Knowledge. Prepared by D.E. Owen [et al.] for U.S. Dept. of Energy. 8p. Bibl.
Research Location: Idaho Falls, Id.: EG and G Idaho, Inc.
Conference: Cambridge, Mass.: Intl. Mting. on LWR Severe Accident Evaluation, 28 August 1983.
Order: NTIS, Order No. DE84000892. Prices: PC A02/MF A01.
Summarizes findings concerning TMI-2 core damage after examination of the damaged reactor core by a remotely controlled camera looking through a leadscrew opening.

632. TMI-2 ACCIDENT RECOVERY ACTIVITIES
Characterization of TMI Type Wastes and Solid Products. Prepared by K.J. Swyler [et al.] for U.S. Dept. of Energy. 12p. Bibl.
Research Location: Upton, N.Y.: Brookhaven National Laboratory.
Conference: Denver, Col.: Ann. Participants Inf. Mting. of DOE Low Level Waste Management Program, 30 August 1983.
Order: NTIS, Order No. DE83018291. Prices: PC A02/MF A01.
Discusses fission products and radioactive wastes resulting from the TMI-2 accident.

633. TMI-2 ACCIDENT RECOVERY ACTIVITIES
Applications of Solid State Track Recorder Neutron Dosimetry for Fuel Debris Location in the TMI Unit 2 Reactor Coolant System. Prepared by F.H. Ruddy [et al.] for U.S. Dept. of Energy. 10p. Bibl.
Research Location: Richland, Wash.: Hanford Engineering Development Laboratory.
Conference: Acapulco, Mexico: 12th Intl. Conference on Solid State Nuclear Track Detectors, 4 September 1983.
Order: NTIS, Order No. DE84006351. Prices: PC A02/MF A01.
Describes use of neutron dosimetry to locate fuel debris in TMI-2's coolant system as an important part of the TMI-2 recovery program.

634. TMI-2 ACCIDENT RECOVERY ACTIVITIES
Acquisition of TMI-2 Core Debris Samples. Prepared by D.E. Owen [et al.] for U.S. Dept. of Energy [et al.]. 6p. Bibl.
Research Location: Idaho Falls, Id.: EG and G Idaho, Inc.
Conference: ANS Winter Mting., 30 October 1983.
Order: NTIS, Order No. DE84002893. Prices: PC A02/MF A01.
Summary of special equipment designed to recover samples from the TMI-2 core debris. Two core debris rubble bed samplers, a sampler deployment boom, radiation shielding equipment and sample casks are described.

635. TMI-2 ACCIDENT RECOVERY ACTIVITIES
Effect of Three Mile Island LOCA on Cables, Connector Components. Prepared by N.S. Cannon [et al.] for U.S. Dept. of Energy. 7p. Bibl.
Research Location: Richland, Wash.: Hanford Engineering Development Laboratory.
Conference: San Francisco, Calif.: ANS Winter Mting., 30 October 1983.
Order: NTIS, Order No. DE84006363. Prices: PC A02/MF A01.
Discusses both in-situ and laboratory tests designed to determine how the different environmental changes during the TMI-2 loss of coolant accident affected the electrical cables and connectors. The different conditions are radiation, containment spray steam, temperature increases, humidity, water pressure, water submersion, and hydrogen fire.

636. TMI-2 ACCIDENT RECOVERY ACTIVITIES
Fuel Debris Assessment for TMI Unit 2 Reactor Recovery by Gamma-Ray and Neutron Dosimetry. Prepared by R. Gold [et al.] for U.S. Dept. of Energy. 17p. Bibl.
Research Location: Richland, Wash.: Hanford Engineering Development Laboratory.
Conference: Zurich, Switzerland: Int'l. Conference on Nondestructive Evaluation in the Nuclear Industry, 27 November 1983.
Order: NTIS, Order No. DE84003950. Prices: PC A02/MF A01.
Describes the location and quantification of fuel debris present in TMI-2's coolant system following the TMI-2 accident.

637. TMI-2 ACCIDENT RECOVERY ACTIVITIES
High Integrity Containers: A Demonstrated Disposal Alternative to Solidification of Radioactive Wastes. Prepared by R.C. Schmitt [et al.] for U.S. Dept. of Energy. 8p. Bibl.
Research Location: Idaho Falls, Id.: EG and G Idaho, Inc.
Conference: Tucson, Az.: Waste Management Conference, 11 March 1984.
Order: NTIS, Order No. DE84011750. Prices: PC A02/MF A01.
Technical description of container designed and tested to provide the safe disposal of EPICOR-II prefilter liners, containing TMI-2 radioactive wastes, without previously solidifying the resins within the containers.

638. TMI-2 ACCIDENT RECOVERY ACTIVITIES
Strippable Coating Used for the TMI-2 Reactor Building Decontamination. Prepared by J.W. Adams [et al.] for U.S. Dept. of Energy. 5p. Bibl.
Research Location: Upton, N.Y.: Brookhaven National Laboratory.
Conference: Tucson, Az.: Waste Management Conference, 11 March 1984.
Order: NTIS, Order No. DE84010516. Prices: PC A02/MF A01.
Summarizes information on strippable coating materials tested for use in decontamination and recovery activities in the TMI-2 reactor building.

639. TMI-2 ACCIDENT RECOVERY ACTIVITIES
Fission Product Release Rates Measured During In-pile Fuel Damage Tests. Prepared by K. Vinijamuri [et al.] for U.S. Dept. of Energy. 8p. Bibl.
Research Location: Idaho Falls, Id.: EG and G Idaho, Inc.
Conference: New Orleans, La.: Ann. Mting. of ANS, 3 June 1984.
Order: NTIS, Order No. DE84014950. Prices: PC A02/MF A01.
Summarizes findings of a series of severe fuel damage tests, performed at the Idaho National Engineering Laboratory and designed to measure the release, transport, and disposition of fission products under conditions simulating the TMI-2 emergency.

640. TMI-2 ACCIDENT RECOVERY ACTIVITIES
Ultrasonic Mapping of the Post Accident TMI-2 Core Configura-

ation. Prepared by L.S. Beller and M.R. Martin for U.S. Dept. of Energy. 5p. Bibl.
Research Location: Idaho Falls, Id.: EG and G Idaho, Inc.
Conference: New Orleans, La.: Ann. Mting. of ANS, 3 June 1984.
Order: NTIS, Order No. DE84014952. Prices: PC A02/MF A01.
   Summarizes the techniques and findings of remote video surveys, made in the summer of 1982, to describe the condition of the TMI-2 reactor core. Techniques used to measure the cavity in the TMI-2 core are described.

641. TMI-2 ACCIDENT RECOVERY ACTIVITIES
Evaluation of Zeolite Mixtures for Decontaminating High-Activity-Level Water at the TMI Unit 2 Nuclear Power Station. Prepared by E.D. Collins [et al.] for U.S. Dept. of Energy [et al.]. 20p. Bibl.
Research Location: Vienna, Austria: IAEA Technical Committee on Inorganic Ion Exchangers and Absorbents for Chemical Processing, 12 June 1984.
Conference: Oak Ridge, Tenn.: Oak Ridge National Laboratory.
Order: NTIS, Order No. DE84012564. Prices: PC A02/MF A01.
   Evaluates effectiveness of various zeolite mixtures, designed for use in decontaminating water which spilled into the TMI-2 auxiliary and fuel handling buildings during the TMI-2 accident.

642. TMI-2 ACCIDENT RECOVERY ACTIVITIES
Radiation Effects on Resins and Zeolites at TMI Unit II. Prepared by J.K. Reilly [et al.] for U.S. NRC [et al.]. 11p. Bibl.
Research Location: Idaho Falls, Id.: EG and G Idaho, Inc.
Conference: Williamsburg, Va.: 12th Intl. Symposium on Effects of Radiation on Materials, 18 June 1984.
Order: NTIS, Order No. DE84015478. Prices: PC A02/MF A01.
   Summarizes experiments which indicate that the EPICOR II liners, containing radioactive waste products similar to those from the damaged TMI-2 reactor, experienced little or no hydrogen generation during two years of storage. Special precautions regarding hydrogen generation should not be necessary for shipment and further storage of EPICOR II liners containing TMI-2 radioactive waste products.

643. TMI-2 ACCIDENT RECOVERY ACTIVITIES
Delayed Neutron Method for Measurement of Fissile/Fertile Content of Samples Ranging from Environmental to Irradiated Fuel.
Prepared by A.E. Proctor [et al.] for U.S. Dept. of Energy. 11p. Bibl.
Research Location: Idaho Falls, Id.: EG and G Idaho, Inc.
Conference: Columbus, Ohio: 25th Ann. Mting. of the Inst. of Nuclear Materials Management, 14 July 1984.
Order: NTIS, Order No. DE84015543. Prices: PC A02/MF A01.
   Discusses the delayed-neutron assay method which is capable of determining the amounts of uranium 235 and uranium 238 in a wide variety of materials. Method is used successfully on samples of TMI-2 reactor core debris.

644. TMI-2 ACCIDENT RECOVERY ACTIVITIES
Radionuclide Distribution in TMI-2 Reactor Building Basement:
Liquids and Solids. Prepared by J.T. Horan [et al.] for U.S. Dept.
of Energy. 12p. Bibl.
Research Location: Idaho Falls, Id.: EG and G Idaho, Inc.
Conference: Snowbird, Vt.: Topical Mting. on Fission Product Behavior and Source Term Research, 15 July 1984.
Order: NTIS, Order No. DE84016216/XAB. Prices: PC A02/MF A01.
Presents sampling collection technique designed to determine the location, quantity and composition of radioactive fission products released into the basement of the TMI-2 reactor building during the TMI-2 emergency. Basement samples are analyzed and described.

645. TMI-2 ACCIDENT RECOVERY ACTIVITIES
Results of Surface Activity and Radiation Field Measurements Made During Surface Decontamination Experiments Conducted at TMI-2.
Prepared by C.V. McIsaac [et al.] for U.S. Dept. of Energy. 16p. Bibl.
Research Location: Idaho Falls, Id.: EG and G Idaho, Inc.
Conference: Snowbird, Vt.: Topical Mting. on Fission Product Behavior and Source Term Research, 15 July 1984.
Order: NTIS, Order No. DE84016217/XAB. Prices: PC A02/MF A01.
Summarizes findings of TMI-2 Gross Decontamination Experiment, carried out in February and March 1982, to evaluate the effectiveness of various surface decontamination techniques inside the TMI-2 reactor building.

646. TMI-2 ACCIDENT RECOVERY ACTIVITIES
TMI-2 Core Debris Analytical Methods and Results. Prepared by D.W. Akers and B.A. Cook for U.S. Dept. of Energy. 11p. Bibl.
Research Location: Idaho Falls, Id.: EG and G Idaho, Inc.
Conference: Snowbird, Vt.: Topical Mting. on Fission Product Behavior and Source Term Research, 15 July 1984.
Order: NTIS, Order No. DE84016108. Prices: PC A02/MF A01.
Summarizes planning and preliminary findings for six samples of TMI-2 core debris carried out in September 1983. Data analysis is presented.

647. TMI-2 ACCIDENT RECOVERY ACTIVITIES
TMI-2 Leadscrew Radionuclide Deposition and Characterization.
Prepared by K. Vinjamuri [et al.] for U.S. Dept. of Energy. 14p. Bibl.
Research Location: Idaho Falls, Id.: EG and G Idaho, Inc.
Conference: Snowbird, Vt.: Topical Mting. on Fission Product Behavior and Source Term Research, 15 July 1984.
Order: NTIS, Order No. DE84016117. Prices: PC A02/MF A01.
Presents detailed chemical and radiological analyses of deposits taken from two TMI-2 reactor leadscrews, HB and BB, following the TMI-2 accident to determine their condition and to estimate conditions within the TMI-2 reactor before decontamination activities can be begun.

Conference Papers

648. TMI-2 ACCIDENT RECOVERY ACTIVITIES
   Applications of Transient Response Characterization and Enhancement to Thermal Instrumentation. Prepared by B.L. Bainbridge and N.R. Keltner for U.S. Dept. of Energy. 19p. Bibl.
   Research Location: Albuquerque, N.M.: Sandia National Laboratories.
   Conference: Knoxville, Tenn.: Industrial Temperature Management Symposium, 10 September 1984.
   Order: NTIS, Order No. DE84015258. Prices: PC A02/MF A01.
   Discusses experiments whereby thermal sensor instruments are utilized to monitor temperatures in the TMI-2 reactor and the containment buildings.

649. TMI-2 ACCIDENT REPORTS
   Tellurium Behavior During and After the TMI-2 Accident. Prepared by K. Vinjamuri [et al.] for U.S. Dept. of Energy. 20p. Bibl.
   Research Location: Idaho Falls, Id.: EG and G Idaho, Inc.
   Conference: Karlsruhe, F.R. Germany: Int'l. Mting. on Thermal Nuclear Safety, 10 September 1984.
   Order: NTIS, Order No. DE85003429/XAB. Prices: PC A02/MF A01.
   An estimate of the probable behavior of tellurium both during and following the TMI-2 accident. Tellurium releases are calculated from available environmental data.

650. TMI-2 ACCIDENT RECOVERY ACTIVITIES
   Advances in Continuous Gamma-ray Spectrometry and Applications. Prepared by R. Gold [et al.] for General Electric, Inc. and U.S. Dept. of Energy. 16p. Bibl.
   Research Location: Richland, Wash.: Hanford Engineering Development Laboratory.
   Conference: Geesthacht, F.R. Ger.: ASTM-Euratom Symposium on Reactor Dosimetry, 24 September 1984.
   Order: NTIS, Order No. DE85003018/XAB. Prices: PC A02/MF A01.
   Describes recent advances and new applications in continuous Compton recoil gamma-ray spectrometry. Applications of fuel distributions are presented relative to the TMI-2 reactor recovery.

651. TMI-2 ACCIDENT RECOVERY ACTIVITIES
   Characterization of Fuel Distributions in the TMI Unit 2 Reactor System by Neutron and Gamma-Ray Dosimetry. Prepared by R. Gold, [et al.] for U.S. Dept. of Energy. 5p. Bibl.
   Research Location: Richland, Wash.: Hanford Engineering Development Laboratory.
   Conference: Geesthacht, F.R. Germany: ASTM-Euratom Symposium on Reactor Dosimetry, 24 September 1984.
   Order: NTIS, Order No. DE85000292/XAB. Prices: PC A02/MF A01.
   Discusses fuel assessments conducted in TMI-2's reactor core and in the primary coolant system to determine the dispersal of fuel debris throughout the TMI-2 reactor system.

652. TMI-1 RESTART
RELAPS/MOD2 Overview and Developmental Assessment Results from TMI-1 Plant Transient Analysis. Prepared by J.C. Lin [et al.] for U.S. Dept. of Energy. 8p. Bibl.
Research Location: Idaho Falls, Id.: EG and G Idaho, Inc.
Conference: Taipei, Taiwan: Intl., Thermal Hydraulics and Plant Operations Topical Mting., 22 October 1984.
Order: NTIS, Order No. DE85003549/XAB. Prices: PC A02/MF A01.
Discribes improvements to the computer code designed to improve safety of TMI-1's reactor operations. Simulation testing indicates highly subcooled water can be safely injected into a high pressure primary coolant system like TMI-1 without problems.

653. TMI-2 ACCIDENT RECOVERY ACTIVITIES
TMI-2 Core Sample Acquisition and Project Examination Status. Prepared by K.C. Sumpter for U.S. Dept. of Energy. 3p. Bibl.
Research Location: Idaho Falls, Id.: EG and G Idaho, Inc.
Conference: Gaithersburg, Md.: 12th Water Reactor Safety Research Information Mting., 23 October 1984.
Order: NTIS, Order No. DE85006378/XAB. Prices: PC A02/MF A01.
Summarizes TMI-2 Core Sample Acquisition and Examination Project, with particular attention to project items obtained and now under examination by various U.S. laboratories. The following items are now being examined: sub-surface core debris samples, resistance thermal detectors, control rod leadscrews, a leadscrew support tube, make-up and purification system filters, and in-situ documentation data.

654. TMI-2 ACCIDENT RECOVERY ACTIVITIES
TMI-2 Plant Demineralizer Sample Analysis. Prepared by J.D. Thompson and J.B. Knauer for ORNL and U.S. Dept. of Energy. 4p. Bibl.
Research Location: Idaho Falls, Id.: EG and G Idaho, Inc.
Conference: Washington, D.C.: Joint Mting. of the ANS and Atomic Industrial Forum, 11 November 1984.
Order: NTIS, Order No. DE85003263/XAB. Prices: PC A02/MF A01.
Describes a proposed sequential elution process, designed to remove cesium 137 from the TMI-2 demineralizer resins.

655. NUCLEAR POWER PLANT SAFETY
Water-Level and Fuel-Failure External Monitoring. Prepared by A. DeVolpi for U.S. Dept. of Energy. 3p. Bibl.
Research Location: Argonne, Ill.: Argonne National Laboratory.
Conference: Washington, D.C.: Symposium on New Technologies in Nuclear Power Plant Instrumentation, 28 November 1984.
Order: NTIS, Order No. DE84011727/XAB. Prices: PC A02/MF A01.
Describes a gamma-ray hodoscope which is designed for use with water cooled reactors and could give plant operators a more accurate knowledge of water level and coolant density in the reactor core. Had such a monitor been available during the early stages of the TMI-2 accident, considerable accident damage could have been avoided.

Conference Papers 197

656. TMI-2 ACCIDENT RECOVERY ACTIVITIES
Comparison of Actual and Predicted Routes Used in the Shipment of Radioactive Materials. Prepared by D.S. Joy [et al.] for University of Tenn. and U.S. Dept. of Energy. 11p. Bibl.
Research Location: Oak Ridge, Tenn.: Oak Ridge National Laboratory.
Conference: Tucson, Az.: Waste Management Conference, 24 March 1985.
Order: NTIS, Order No. DE85011558/XAB. Prices: PC A02/MF A01.
Summarizes factors involved in selection of highway routes for shipment of TMI-2 radioactive materials made from TMI to Idaho test sites in 1982 and 1983. Various factors influenced choice of routes: highway conditions; distance; efforts to avoid urban areas with population of 100,000 or more; weather conditions; and driver preference.

657. TMI-2 ACCIDENT ACTIVITIES
Degradation of Resins in EPICOR-II Prefilters from TMI. Prepared by J.W. McConnell and R.D. Sanders for U.S. Dept. of Energy. 8p. Bibl.
Research Location: Idaho Falls, Id.: EG and G Idaho, Inc.
Conference: Tucson, Az.: Waste Management Conference, 24 March 1985.
Order: NTIS, Order No. DE85008582/XAB. Prices: PC A02/MF A01.
Examines data concerning chemical and physical condition of synthetic ion exchange resins found in EPICOR-II prefilters PF-8 and -20 to determine possible causes for degradation.

658. TMI-2 ACCIDENT RECOVERY ACTIVITIES
Institutional Agreements and Interaction for Research and Disposition of Nuclear Waste from TMI. Prepared by J.K. Reilly [et al.] for U.S. NRC and U.S. Dept. of Energy. 2p. Bibl.
Research Location: Washington, D.C.: U.S. Dept. of Energy.
Conference: Tucson, Az.: Waste Management Conference, 24 March 1985.
Order: NTIS, Order No. DE85008929/XAB. Prices: PC A02.
Summarizes the various institutional agreements made with respect to research and disposition of nuclear wastes from the 1979 emergency.

659. TMI-2 ACCIDENT RECOVERY ACTIVITIES
Metallographic Examination of EPICOR-II Liners from TMI.
Prepared by J.W. McConnell, Jr., H.W. Spaletta for U.S. Dept. of Energy. 7p. Bibl.
Research Location: Idaho Falls, Id.: EG and G Idaho, Inc.
Conference: Tucson, Az.: Waste Management Conference, 24 March 1985.
Order: NTIS, Order No. DE85008651/XAB. Prices: PC A02/MF A01.
Describes examination of selected samples from TMI-2 EPICOR-II prefilter liners to make certain that the liners and their contents can be safely stored at the Idaho National Engineering Laboratory

for ten years. Examination results indicate the liners have a fifty-year lifespan.

660. TMI-2 ACCIDENT RECOVERY ACTIVITIES
Processing and Removal of the TMI Makeup and Purification System Resins. Prepared by J.K. Reilly [et al.] for U.S. Dept. of Energy [et al.]. 6p. Bibl.
Research Location: Idaho Falls, Id.: Eg and G Idaho, Inc.
Conference: Tucson, Az.: Waste Management Conference, 24 March 1985.
Order: NTIS, Order No. DE85008928/XAB. Prices: PC A02.
Describes cooperative efforts between GPU Nuclear, Inc. and the U.S. Dept. of Energy and others to develop specialized equipment and chemical processing methods to elute radioactive cesium from the TMI-2 makeup and purification demineralizers.

661. TMI-2 ACCIDENT RECOVERY ACTIVITIES
TMI-2 Spent Fuel Shipping. Prepared by G.J. Quinn and H.M. Burton for U.S. Dept. of Energy. 3p. Bibl.
Research Location: Idaho Falls, Id.: EG and G Idaho, Inc.
Conference: Tucson, Az.: Waste Management Conference, 24 March 1985.
Order: NTIS, Order No. DE85009972/XAB. Prices: PC A02/MF A01.
Describes details of proposed railroad shipment of TMI-2 failed fuel from TMI to the Idaho National Engineering Laboratory. Shipping canister design and design for shipping casks, accommodating seven individual canisters of damaged fuel, are described.

662. TMI-2 ACCIDENT RECOVERY ACTIVITIES
Cleanup of TMI-2 Demineralizer Resins. Prepared by W.D. Bond [et al.] for U.S. Dept. of Energy [et al.]. 27p. Bibl.
Research Location: Idaho Falls, Id.: EG and G Idaho, Inc.
Conference: Miami, Fla.: 189th National Mting. of ACS, 28 April 1985.
Order: NTIS, Order No. DE85011431/XAB. Prices: PC A03/MF A01.
Describes technical process whereby radiocesium can be separated from the TMI-2 demineralizer resins and then can be reabsorbed onto the zeolite ion exchangers in the TMI-2 Submerged Demineralizer System (SDS).

663. TMI-2 ACCIDENT RECOVERY ACTIVITIES
Development of the Flowsheet Used for Decontaminating High-Activity-Level Water at TMI-2. Draft. Prepared by E.D. Collins [et al.] for U.S. Dept. of Energy [et al.]. 31p. Bibl.
Research Location: Idaho Falls, Id.: EG and G Idaho, Inc.
Conference: Miami, Fla.: 189th National Mting. of ACS, 28 April 1985.
Order: NTIS, Order No. DE85012782/XAB. Prices: PC A03/MF A01.
Describes development of an improved chemical processing flowsheet, to be used in decontamination of TMI-2's radioactive water. The new process will improve the radioactive waste processing and will render the wastes into a suitable form for disposal.

Conference Papers

664. TMI-2 ACCIDENT REPORTS
Thermal Hydraulic Features of the TMI Accident. Prepared by B. Tolman for U.S. Dept. of Energy. 28p. Bibl.
Research Location: Idaho Falls, Id.: Idaho National Engineering Laboratory.
Conference: Miami, Fla.: Ann. Mting. of ACS, 1 May 1985.
Order: NTIS, Order No. DE85014569/XAB. Prices: PC A03/MF A01.
Suggests TMI accident data will have great value in assessing possible consequences of any future severe core damage nuclear power plant accidents and in developing additional plant safety procedures.

665. TMI-2 ACCIDENT RECOVERY ACTIVITIES
Role of Ion Exchange in the TMI Cleanup. Prepared by R.M. Wallace for U.S. Dept. of Energy. 13p. Bibl.
Research Location: Aiken, S.C.: DuPont Savannah River Laboratory.
Conference: Harwell, U.K.: Intl. Mting. on Solvent Extraction and Ion Exchange in the Nuclear Cycle, 3 September 1985.
Order: NTIS, Order No. DE85008259/XAB.
Summarizes details of how contaminated water, found in three locations following the TMI-2 emergency, was decontaminated through use of two different ion exchange systems, the EPICOR II System and the Submerged Demineralizer System.

666. TMI-2 ACCIDENT RECOVERY ACTIVITIES
Neutron Dosimetry in the TMI Unit 2 Reactor Cavity with Solid State Track Recorders. Prepared by R. Gold [et al.] for U.S. Dept. of Energy. 5p. Bibl.
Research Location: Richland, Wash.: Hanford Engineering Development Laboratory.
Conference: Rome, Italy: 13th Intl. Conference on Solid State Nuclear Track Detectors, 23 September 1985.
Order: NTIS, Order No. DE85018470/XAB. Prices: PC A02/MF A01.
Summarizes solid-state track recorder neutron dosimetry findings concerning probable radiation streaming in the damaged TMI-2 reactor cavity. Estimates indicate there are probably at least two tons of fuel lying at the bottom of the TMI-2 reactor vessel.

667. NUCLEAR POWER PLANT SAFETY
Integral Systems Test (IST) Program Facility Scaling and Integration. Prepared by T.K. Larson for U.S. Dept. of Energy. 41p. Bibl.
Research Location: Washington, D.C.: U.S. Nuclear Regulatory Commission.
Conference: Gaithersburg, Md.: 13th Information Mting. of Water Reactor Safety Research, 22 October 1985.
Order: NTIS, Order No. DE86003220/XAB. Prices: PC A03/MF A01.
Describes cooperative program initiated by the Electric Power Research Institute and the U.S. Nuclear Regulatory Commission. The program is designed to study various unresolved safety issues raised by the TMI-2 emergency and uses three test sites, each of which has different operating capabilities and a Babcock and Wilcox nuclear steam supply system.

668. TMI-2 ACCIDENT REPORTS
Tellurium Chemistry, Tellurium Release and Disposition During the TMI-2 Accident. Prepared by K. Vinjamuri [et al.] for Sandia National Laboratories and U.S. Dept. of Energy. 23p. Bibl. Research Location: Idaho Falls, Id.: EG and G Idaho, Inc. Conference: Gaithersburg, Md.: 13th Information Mting. on Water Reactor Safety Research, 22 October 1985. Order: NTIS, Order No. DE86003224/XAB. Prices: PC A02/MF A01.
Discusses tellurium chemistry and the probable behavior of tellurium both during and after the TMI-2 accident. Research and test samples indicate only small amounts of tellurium were released and transported from TMI-2's reactor core during the emergency, probably due to presence of zircalloy cladding and other structural materials.

NAME INDEX

(Includes Authors, Editors and Principal Investigators)

Please note: Numbers refer to record numbers, not to pages.

Abbott, W. H. 548
Abramson, P. B. 370, 379
Adams, J.W. 638
Agnew, Harold 282
Akers, D. W. 530, 531, 646
Ales, M. W. 506
Allen, R. P. 566
Allison, C. M. 606
Alvares, N. J. 432, 496, 614
Alvarez, J. L. 520, 592, 608, 626
Appel, J. N. 550
Armento, W. J. 126, 580
Arrowsmith, H.W. 566
Ayers, A. L., Jr. 500, 527, 528

Bainbridge, B. L. 648
Bandy, R. 223
Baretta, A. J. 378
Barefoot, E. D. 425
Barletta, R. E. 227, 605
Barner, J. O. 543
Baston, V. 538
Bassett, B. 596
Battist, Lewis 56
Beers, R. H. 498
Behling, S. R. 544
Beller, L. S. 387, 501, 641
Bengal, Paul R. 540, 559
Bennett, P. R. 555
Berger, C. D. 362, 363, 392
Berman, Marshall 166, 303
Bibler, N. E. 586

Bixby, W. W. 607
Bond, W. D. 662
Bores, R. J. 167
Boudreau, Jay E. 280
Bower, J. M. 415
Bradley, William 278
Breen, R. L. 283
Bretthauer, Erich W. 400, 401, 402, 403, 404, 405, 406
Bromet, Evelyn 11
Brooksbank, R. E. 126, 353, 513, 560, 576, 579
Brown, H. L. 501
Bryan, G. H. 450, 512, 525
Buelt, J. L. 381, 420, 448
Bulkin, Barb 282
Burns, R. D., III 571
Burton, H. M. 591, 607, 661

Calloway, N. E. 385
Campbell, D. O. 126, 436
Cannon, N. S. 635
Cantelon, Philip L. 357
Card, C. J. 477
Carew, J. F. 569
Carlson, J. O. 532
Carne, Alan T. 337
Carter, G. S. 431
Castro, William R. 246
Caswell, V. D. 351
Chalmers, J. A. 126
Chang, Y. I. 389
Chapman, Gordon 118
Chapman, R. L. 461, 603

Chenault, William W. 359
Chester, C. V. 511
Civiak, Robert L. 118, 298
Clark, Gene 251
Clark, R. L. 536
Cline, J. E. 388
Cole, Randall K., Jr. 44
Collins, Daniel L. 516
Collins, E. D. 578, 587, 599, 641, 663
Colton, D. P. 452
Cook, B. A. 646
Cotter, S. J. 358
Cottrell, William B. 246, 249, 254, 258, 261, 290, 291
Cox, T. E. 437, 468, 484, 612
Croucher, D. W. 408, 585
Cummings, G. E. 564
Cummings, John C. 303
Cunningham, George, W. 266, 272

Daniel, J. L. 449
Davis, R. E. 227
Davis, R. J. 519
Davis, W. Kenneth 293
DeVolpi, A. 655
Dickerson, M. H. 361, 561
Digon, Edward 16
DiSabella, Renee 24
Distenfield, C. 563
Dodge, R. L. 604
Doerge, D. H. 220
Donnelly, Warren, H. 107
Dornsife, William P. 312
Downing, R. H. 92
Doyle, J. D. 463
Doyle, T. E. 520, 608, 626
Dress, W. B. 214
Dreyer, N. A. 100
Dynes, Russell, R. 59

Eidam, G. E. 397, 611
Ellis, W. D. 545
England, T. R. 352, 371, 491
Ettinger, B. A. 620
Evans, D. L. 421

Fabrikant, Jacob I. 64, 574
Falk, D. E. 558
Feeley, J. J. 584
Fienberg, Stephen E. 341
Firebaugh, Morris W. 274
First, Melvin W. 289
Fisher, H. L. 351
Flynn, Cynthia B. 84, 136
Flynn, J. 193
Folkenberg, Judy 322
Franz, W. A. 621
Fritzsche, A. E. 350
Fryer, M. O. 453

Gallucci, Raymond H. V. 440
Gamble, H. B. 92
Gannon, J. A. 416, 466
Gardner, H. R. 510, 542
Garner, R. W. 458
Gilbert, Humphrey 289
Golay, Michael W. 323
Gold, R. 556, 601, 636, 650, 651, 666
Goldhaber, Marilyn 24, 27, 29, 301
Goldman, M.I. 482, 483
Gorinson, Stanley M. 60, 61, 62
Goris, P. 373
Gotchy, R. L. 167
Greenborg, J. 610
Greenlee, D. W. 386
Gudiksen, P. K. 561

Hagen, Edward W. 275, 292
Hamilton, W. H. 488
Hansen, R. F. 497
Hardy, R. 552
Hartwell, J. K. 376
Harvego, E. A. 276
Hatcher, James C. 109
Hecker, L. A. 600
Heintzleman, R. E. 419
Helbert, H. J. 529
Henrie, J. O. 454, 455, 480, 518, 550, 595
Hetzer, D. C. 619
Hickey, C. R., Jr. 49, 163
Higson, Donald J. 328

Name Index

Hobbins, R. R. 630
Hoenes, G. R. 412
Holtzworth, R. E. 590
Horan, J. Thomas 148, 644
Houts, Peter S. 5, 24
Hsu, C. J. 575
Hu, Teh-wei 6, 18
Hull, A. P. 562

Ireland, John R. 281, 285, 577
Israel, S. 565

Jackson, James F. 279
Jacoby, J. K. 460
Jansen, Steven D. 391
Jenkins, W. W. 476
Johnson, C. E. 370
Johnson, D. A. 530, 531
Johnson, S. J. 592
Jones, J. E. 382, 383, 422, 423, 424, 425, 426, 427, 428, 429, 430, 438
Jones, R. H. 583
Joy, D. S. 656

Kalman, George 304
Kanapilly, G. M. 375
Keefer, D. G. 539
Kelly, K. 223
Keltner, N. R. 648
Kemeny, John 104
Kim, S. S. 502
King, L. J. 513, 625
Kirchner, W. L. 573
Klemish, J. 563
Knauer, J. B. 433, 654
Knox, J. B. 372
Kovach, J. Louis 287
Kripps, L. T. 407
Kudera, D. E. 628

Larson, T. K. 276, 355, 667
Lauer, G. B. 509
Lehman, James E. 29
Leidam, Gregory R. 148
Leigh, D. W. 611

Leverett, Miles C. 273
Levine, Saul 270
Lin, J. C. 652
Lo, Ronnie 114
Logan, J. A. 622
Long, A. B. 259, 315
Lorenz, R. A. 465

McConnell, J. W., Jr. 233, 464, 515, 546, 657, 659
McIsaac, C. V. 487, 521, 539, 619, 645
McLaughlin, I. B. 447, 490
Malloy, D. J. 389
Malone, T. B. 105
Manning, D. T. 294
Manno, Vincent P. 323
Marrone, Joseph 299
Mason, R. E. 471, 473
Mast, P. K. 568
Mathis, M. V. 438
Maudlin, P. J. 629
Maxey, Margaret N. 284
Mays, Gary T. 275, 329
Meade, J. P. 545
Meinhold, C. B. 75
Milinauskas, A. P. 451
Miller, C. W. 570
Miller, Kevin M. 57
Miller, R. L. 220, 541
Miller, W. J. 364
Mock, J. W. 439
Mohanty, A. K. 399
Moore, J. A. 472
Moore, Mickey M. 31
Morrison, J. L., Jr. 398
Mueller, G. M. 418, 467
Muhlheim, M. D. 317, 318, 321, 330
Munson, L. F. 210, 552
Muraka, Ishwar P. 347
Murphy, M. B. 380, 419, 594
Mynatt, F. R. 362, 363

Nagata, P. K. 414
Neilson, R. M., Jr. 523
Newman, L. W. 493
Newmann, R. C. 597, 627
Nitschke, R. L. 410

Ogle, R. E. 618
Osterhoudt, T. R. 537
Owen, D. E. 598, 602, 631, 634

Palladino, Nunzio J. 288, 324
Parkinson, Davis 11
Parlee, K. 609
Pasupathi, V. 459
Pellitier, C. A. 478, 616
Pentecost, Edwin D. 347
Perham, Christine 286
Perry, Ronald W. 377
Peterson, Russell, W. 248
Phillips, J. R. 474
Phung, Doan L. 311, 499, 522
Polentz, L. M. 542
Porri, Willis P. 329
Proctor, A. E. 643
Postma, A. K. 454, 455, 595
Pryor, Richard J. 277

Queen, S. P. 475
Quinn, G. J. 486, 517, 524, 554, 661

Rados, Bill 247
Raleigh, H. D. 345
Rankey, E. J. 476
Reilly, J. K. 642, 658, 660
Reisch, Frioyes 329
Reno, H. W. 461, 604
Rest, J. 370, 379
Rich, B. L. 367, 384
Rikli, Patricia 334
Robert, Jack O. 127
Rossi, Harold H. 260
Ruddy, F. H. 441, 633
Runion, T. C. 589
Russ, E. K. 298

Saltman, Jerome 127
Samworth, R. B. 49
Sanchez, H. F. 486
Sanders, R. D., Sr. 233
Sargent, Marilyn 331
Scherpeiz, R. I. 495

Schmitt, R. C. 617, 637
Schultz, H. W. 414
Schwarz, W. F. 613
Schwegler, E. C. 629
Scott, D. D. 373
Sears, Gerald E. 106
Shrivastava, P. K. 109
Shuping, Ralph E. 175
Siemens, D. H. 588
Silver, E. G. 295, 296, 300, 305, 306, 307, 313, 314, 317, 318, 327, 330, 332, 333, 336, 338
Slaysman, Kenneth S. 18
Smith, James G. 342, 343
Snyder, Bernard J. 114
Soberano, F. T. 442, 446, 507, 508
Solomon, Kenneth A. 366
Spaletta, H. W. 515
Stevenson, Michael O. 279
Strahm, R. C. 494
Streufert, Siegfied 44
Sumpter, K. C. 615, 653
Sumstine, R. L. 489
Swenson, C. E. 558
Swyler, K. J. 632

Thoma, John O. 217
Thome, F. V. 594
Thomas, J. T. 434, 503
Thompson, J. D. 654
Thompson, Loren 310
Tokuhata, George 10, 16, 20, 23, 25
Tolman, B. 664
Thornburg, Dick (Gov.) 8, 17, 325
Towers, D. A. 178
Townes, G. A. 374
Trauger, D. B. 368, 467

Vigil, John C. 277, 624
Vigil, M. G. 395
Vinjamuri, K. 526, 549, 553, 557, 639, 647, 649, 668
Viskanta, R. 399
Voilleque, P. G. 417, 462

Name Index

Walker, Pamela 192
Wallace, R. M. 665
Watkins, D. E. 485
Welch, J. M. 623
Weller, Richard A. 304
West, J. T. 581
Westfall, R. M. 41, 582
Wilde, N. 398, 479
Wilkins, D. E. 492, 533
Williams, Robert C. 357
Wilson, W. B. 352, 371, 491
Wooley, R. L. 572
Wooton, R. O. 85

Wright, James L., Jr. 12
Wynhoff, N. L. 459

Yancey, M. E. 479, 494, 514, 535
Yesso, J. D. 411

Zebroski, Edwin L. 273
Zeile, H. J. 593

SUBJECT INDEX

Please note: Numbers refer to record numbers, not to pages.

Accident causes and development, TMI-2 63, 184, 203, 225, 247, 249, 255, 271, 277, 352, 455, 567, 584, 629
Accident reports, TMI-2 3, 8, 12, 35, 36, 37, 44, 46, 60, 61, 62, 63, 65, 66, 67, 71, 72, 73, 85, 103, 115, 116, 124, 128, 129, 130, 131, 145, 175, 184, 246, 249, 255, 262, 263, 264, 271, 285, 338, 357, 364, 368, 370, 379, 398, 399, 451, 455, 498, 526, 557, 567, 568, 569, 573, 575, 577, 581, 595, 599, 628, 629, 664, 668
Accident simulation, TMI-2 379, 577, 581, 606
Advertising/public relations campaign 356
Air incinerator design, TMI-2 413
Air-sampling system, TMI-2 563
Asymmetric Multiple Position Neutron Source (AMPNS) Method 502
Atmospheric Release Advisory Capability (ARAC) Method 372, 561
Atomic Energy Commission (U.S.) 31

Babcock and Wilcox Corp. 68, 217, 469, 470, 548, 667
Bechtel Northern Corp. 481
Bibliographies, TMI-2, related reports 7, 456

Cesium releases 155, 433, 519, 530, 531, 534, 542, 586, 654, 660, 662
Civil defense reports, TMI-2 38, 39
Clean-up and recovery, TMI-2 41, 69, 74, 87, 88, 112, 114, 116, 124, 126, 127, 135, 139, 147, 150, 151, 152, 155, 156, 157, 158, 160, 162, 172, 176, 177, 181, 185, 188, 189, 197, 198, 200, 203, 204, 207, 209, 215, 219, 220, 221, 225, 226, 227, 228, 229, 233, 235, 236, 276, 277, 280, 291, 296, 302, 304, 305, 308, 309, 314, 318, 319, 335, 353, 360, 365, 369, 371, 373, 374, 381, 382, 383, 384, 385, 390, 408, 409, 410, 411, 413, 415, 416, 417, 418, 419, 420, 422, 423, 424, 425, 426, 427, 428, 429, 430, 431, 433, 434, 435, 436, 437, 438, 439, 440, 441, 442, 443, 445, 446, 447, 448, 449, 450, 458, 459, 461, 463, 464, 466, 468, 471, 472, 473, 474, 475, 476, 477, 478, 479, 480, 481, 482, 483, 484, 485, 486, 487, 488, 489,

490, 491, 492, 493, 494, 496, 497, 501, 502, 503, 504, 505, 506,
507, 510, 512, 513, 514, 515, 517, 518, 520, 521, 529, 535,
536, 537, 538, 539, 542, 543, 544, 545, 552, 560, 566, 576,
578, 582, 592, 599, 601, 607, 608, 612, 620, 647, 657,
665
Computer code calculations   41, 501, 544, 573, 575, 581, 606, 624,
628, 652
Containment building decontamination, TMI-2   102, 105, 148, 210,
433, 435, 437, 448, 453, 460, 463, 468, 471, 487, 488, 495,
496, 519, 520, 521, 529, 539, 576, 586, 589, 611, 619, 538
Containment building entries, TMI-2   166, 396, 397, 534, 614, 638
Control room design, TMI-1   91, 149
Control room design, TMI-2   105, 292
Core Sample Acquisition and Examination Project, TMI-2   653
Cumberland County (PA)   13
Cumberland County (PA), Office of Emergency Preparedness   13

Dauphin County (PA)   14
Dauphin County (PA), radiological emergency response plan   14
Debris canister design, TMI-2   221, 374, 395, 461, 476, 492, 533,
546, 550, 590, 603, 617, 618, 637, 657, 659
Debris Defueling Working Group, TMI-2   518
Demineralizers, measurement of TMI-2   474
Department of Agriculture (PA)   9
Department of Energy (U.S.)   47, 57, 95, 96, 97, 110, 111, 146,
148, 150, 198, 201, 228, 357, 360, 372, 445, 494, 498, 518,
561, 562, 590, 607, 620, 660
Department of Health (PA)   5, 7, 10, 19, 23, 24, 25, 27
Department of Health (PA), Bureau of Radiological Health   562
Department of Health (PA), Division of Epidemiological Research   10

Economic effects, TMI-2   4, 6, 9, 18, 92, 193, 356
Electric Power Research Institute   667
Electric utility rates   187, 251
Emergency communications systems   38, 65, 164, 170, 178, 259, 362,
363, 593
Emergency evacuation, TMI area residents   3, 29
Emergency Management Agency (PA)   15
Emergency preparedness planning   13, 14, 15, 21, 22, 26, 28, 31,
38, 39, 42, 52, 58, 59, 60, 93, 123, 134, 137, 168, 169, 181,
190, 191, 195, 199, 232, 239, 247, 266, 267, 312, 315, 359,
362, 363, 377, 511, 522, 545, 564, 565, 580, 593
Emergency preparedness planning, Pennsylvania counties   13, 14,
21, 22, 28
Energy sources, alternative   55
Enrico Fermi power plant accident   48
Environmental effects, TMI-2   49, 99, 102, 163, 194, 211, 229, 268,
280, 347, 348, 392, 400, 401, 402, 403, 404, 405, 407, 570,
635

Subject Index   209

Environmental impact statements, TMI Nuclear Station   30, 211, 229,
    344
Environmental Protection Agency (U.S.)   268, 286, 291
EPICOR II system   227, 233, 394, 411, 415, 420, 459, 461, 463, 464,
    475, 476, 505, 515, 523, 546, 589, 604, 605, 617, 618, 637,
    642, 657, 659, 665
Extraordinary Nuclear Occurrence (ENO)   130

Federal Radiation Protection Act of 1979   101
Financial problems, TMI-2   6, 18, 77, 92, 127, 140, 146, 157, 158,
    161, 162, 176, 179, 180, 185, 187, 188, 196, 206, 251, 293,
    330, 332, 354, 490
Fisheries, TMI-2 area   49
Fission products releases, measurement of   491, 509, 585, 612, 632,
    639

General Assembly (PA)   12, 26
General Public Utilities, Inc. (GPU)   2, 62, 127, 173, 176, 207,
    469, 470
General Public Utilities Nuclear, Inc. (GPUN)   661
Generator cracking, TMI-1   597
Governor's Commission on Three Mile Island (PA)   1, 7, 8
Governor's Office of Policy and Planning (PA)   4, 9

Hart Committee report   116, 117
Health-related studies, TMI-2   5, 6, 7, 10, 16, 18, 19, 20, 23, 27,
    47, 64, 70, 72, 76, 100, 101, 106, 109, 110, 145, 167, 245,
    250, 260, 287, 301, 333, 341, 358, 362, 363, 367, 570, 571,
    574
Hydrogen burn   see   Hydrogen combustion
Hydrogen combustion   166, 303, 310, 323, 414, 432, 453, 454, 460,
    488, 496, 536, 538, 595, 614
Hydrogen, generation of   44, 128, 323, 550, 572, 575, 585, 642
Hydrolasing   611
Hypothyroidism, congenital   16, 20

Idaho National Engineering Laboratory   335, 384, 421, 447, 464, 485,
    490, 492, 500, 515, 517, 524, 527, 533, 534, 604, 606, 617,
    618, 639, 659, 661
Institute of Nuclear Power Operations (INPO)   125, 253, 297
Institutional research agreements   658
Integral Systems Test Program   667
International Electrotechnical Commission   320

Jersey Central Power and Light Co.   30, 31, 69

Kemeny Commission see President's Commission on the Accident at
  Three Mile Island (TMI)
KENO-Va Improved Monte Carlo Criticality Program 503
Krypton-85, venting of 75, 86, 99, 126, 139, 361, 376, 388, 407,
  465, 498, 547, 591

Lancaster County (PA) 21
Lancaster County (PA), radiological emergency response plan 21
Lawsuit review, TMI-1 related 469, 470
Lebanon County (PA) 22
Lebanon County (PA), radiological emergency response plan 21
Leadscrew samples, TMI-2, examination of 488, 536, 549, 621, 647
Legal effects, TMI-2 accident 1, 165, 256, 280, 299, 305, 316
Light water reactor safety 68, 364, 365, 366, 478, 499, 522
Los Alamos National Laboratory 568, 624

Maps, TMI-2 related 90, 501, 640
Metropolitan Edison Co. 2, 30, 31, 32, 69, 91, 93, 123, 143, 144,
  149, 316, 349, 354, 367
Mobile Accident Capability Program, TMI-2 457, 562
Mobile response emergency equipment, TMI-2 545
Molecular Optical Laser Examiner (MOLE) TMI-2 520, 592, 608
Mortality, fetal 16
Mortality, infant 16, 29

National Energy Research Institute 17
National Guard (PA) 3
National Technology Foundation 104
National Weather Service (U.S.) 242
Neutron detector measuring system 398
Nuclear Property Insurance Act of 1981 187
Nuclear Regulatory Commission (NRC) 2, 33, 40, 43, 44, 46, 49,
  52, 53, 61, 66, 68, 75, 77, 78, 79, 80, 81, 82, 83, 84, 85,
  86, 87, 88, 89, 91, 92, 93, 94, 97, 108, 112, 113, 114, 115,
  121, 124, 128, 130, 133, 134, 136, 137, 141, 171, 173, 181,
  183, 189, 191, 217, 219, 224, 225, 226, 231, 237, 238, 254,
  255, 256, 288, 305, 339, 499, 551, 667
Nuclear Regulatory Commission (NRC), NRC Action Plan 89, 112,
  114, 135
Nuclear Regulatory Commission (NRC), reorganization proposal 132
Nuclear Safety Analysis Center 273
Nuclear Safety Research and Development act of 1980 122

Oak Ridge National Laboratory 497, 513, 560
Once-through Steam Transient Generator, TMI-2 575
Operating license, TMI-1 346
Operating license, TMI-2 33, 349

Subject Index 211

Operator performance, TMI-2, accident 105
Operator training, power plant operations 105, 522, 564, 584, 593

Pacific Northwest Laboratory 525
Pennsylvania Bureau of Radiological Health see Department of Health (PA), Bureau of Radiological Health
Pennsylvania Department of Agriculture see Department of Agriculture (PA)
Pennsylvania Department of Health see Department of Health (PA)
Pennsylvania Department of Health, Division of Epidemiological Research see Department of Health (PA), Division of Epidemiological Research
Pennsylvania Electric Co. 30, 31, 69
Pennsylvania Emergency Management Agency see Emergency Management Agency (PA)
Pennsylvania General Assembly see General Assembly (PA)
Pennsylvania Governor's Commission on Three Mile Island see Governor's Commission on Three Mile Island (PA)
Pennsylvania Governor's Office of Policy and Planning see Governor's Office of Policy and Planning (PA)
Pennsylvania National Guard see National Guard (PA)
Pennsylvania State University 6
Photographs, atmosphere, TMI area 175
Photographs, TMI-2 containment building 148
Pipe and tube cracking, TMI-1 583, 597
Plant and animal health, TMI-2 area 106
Plenum assembly, damage to 488, 506, 548, 549, 602, 621
Polar crane, development of 220, 302, 307, 318
Population changes, TMI region 24
Population Registry, TMI region 27, 29, 301
Potassium iodide, uses of 26
Power plant construction, decline in starts 248, 257
Power plant design, production of 48, 292
Power plant insurance 177, 179, 187
Power plant, licensing 89, 98, 113, 142, 165, 208, 209, 232, 251, 258, 261, 262, 265, 288, 293
Power plant, regulatory orders 54, 142, 251, 257, 258, 261, 262, 265, 266, 267, 288, 293, 320, 324, 329, 522, 551, 565
Power plant safety and research 26, 37, 40, 45, 48, 50, 51, 52, 94, 95, 96, 97, 103, 104, 118, 119, 121, 122, 125, 138, 142, 159, 160, 182, 183, 186, 190, 191, 198, 200, 210, 217, 218, 224, 231, 236, 238, 239, 248, 252, 254, 259, 265, 269, 270, 272, 273, 275, 278, 279, 281, 282, 283, 289, 292, 298, 311, 312, 315, 320, 328, 337, 364, 366, 434, 454, 457, 499, 509, 522, 551, 565, 579, 580, 584, 593, 607, 620, 624, 655, 664, 667
Power plant, shutdowns 51
Power plant, site selection, considerations for 54
Power plant, waste products, disposal of 54
Pregnancy outcome 16, 20, 23, 25

President's Commission on the Accident at Three Mile Island (TMI)
 34, 53, 59, 60, 61, 62, 63, 64, 65, 67, 108, 129, 131, 240,
 263, 352, 371, 392, 400, 401, 402, 403, 404, 405, 406, 513
Price-Anderson Act   299
Property values and sales, residential, TMI region   92
Psychological stress, induced by TMI-2 accident   11, 84, 192, 322,
 331, 341, 444, 516, 574
Public opinion, nuclear plant safety since TMI-2   393
Public reaction, TMI-2 accident   107, 244, 274, 294, 356, 391, 393,
 571
Purification demineralizers, TMI-2   473, 476, 477, 612, 660, 662
Pyrophoricity studies   436, 536, 538

Quick look examination techniques, TMI-2   621

Radiation exposure studies, TMI-2 related   47, 70, 100, 101, 109,
 110, 367, 372, 570
Radiation monitoring, TMI-2   47, 58, 241, 242, 243, 247, 268, 284,
 286, 350, 378, 384, 495, 498, 509, 547, 560, 561, 562, 563,
 569, 570
Radiation releases, TMI nuclear facility   57, 76, 86, 126, 128, 175,
 194, 199, 202, 229, 241, 242, 243, 247, 260, 268, 286, 334,
 350, 351, 352, 353, 355, 358, 361, 362, 363, 367, 370, 372,
 373, 378, 380, 384, 392, 400, 401, 402, 403, 404, 405, 406,
 407, 412, 418, 419, 422, 423, 424, 425, 426, 427, 428, 429,
 430, 435, 452, 462, 465, 467, 495, 498, 507, 514, 519, 530,
 531, 547, 560, 561, 562, 563, 569, 570, 571, 572, 586, 591,
 628, 632, 635, 639
Radioactive wastes, TMI-2 (liquid and solids), decontamination   102,
 122, 198, 211, 221, 229, 296, 371, 373, 374, 394, 395, 410,
 411, 413, 420, 421, 433, 440, 447, 448, 449, 459, 461, 464,
 474, 476, 477, 497, 505, 512, 513, 515, 523, 524, 527, 528,
 538, 539, 546, 550, 552, 560, 566, 572, 576, 578, 579, 587,
 598, 603, 604, 605, 610, 617, 618, 625, 632, 637, 641, 654,
 658, 660, 663
Radioactive wastes, TMI-2, shipping of   323, 475, 485, 488, 500,
 517, 554, 611, 656, 661
Radioactive wastes, TMI-2, storage of   642
Radiation survey, aerial, TMI-2   175, 452, 498
Radioiodine releases, TMI-2, measurement of   287, 487, 519, 563,
 616
Radionuclide studies, TMI-2   410, 417, 437, 462, 482, 483, 484, 519,
 644, 647
Reactor Building Gross Decontamination Experiment, TMI-2   487,
 619
Reactor cables, TMI-2, examination of   489, 529, 534, 535, 635
Reactor coolant system, TMI-2   410, 473, 497, 519, 633, 636
Reactor core damage, TMI-2   41, 43, 128, 479, 480, 493, 501, 502,
 520, 530, 531, 532, 534, 535, 550, 553, 556, 568, 581, 582,

585, 601, 602, 606, 608, 615, 621, 622, 630, 631, 634, 635, 640, 643, 646, 650, 651, 653, 664, 666
Reactor core, defueling, TMI-2   304, 307, 340, 365, 373, 374, 385, 386, 387, 390, 408, 416, 417, 421, 431, 441, 462, 488, 490, 491, 500, 501, 510, 517, 518, 520, 530, 531, 532, 534, 536, 538, 540, 541, 542, 548, 550, 553, 554, 558, 573, 601, 602, 615, 622, 623, 629, 636, 639, 643, 650, 651, 653
Reactor core, head removal, TMI-2   308, 326, 431, 432, 536, 540, 559, 622
Reactor core, storage of, TMI-2   500, 623
Reactor design, changes in   522, 551
Reactor gauge, safety improvements   205, 214, 655
Reactor heat, transfer of, TMI-2 accident   399
Reactor instruments, TMI-2, examination of   380, 382, 383, 384, 389, 418, 419, 422, 423, 424, 425, 426, 427, 428, 429, 430, 438, 439, 442, 443, 446, 453, 455, 458, 466, 467, 479, 481, 489, 494, 504, 507, 508, 514, 535, 536, 542, 549, 553, 555, 569, 594, 600, 609, 613, 626, 635, 647, 648
Reactor safety, operational   166, 214, 217, 220, 269, 279, 282, 292, 328, 337, 602, 655
Recovery program, TMI-2, Japanese participation in   309
Remote controlled emergency equipment, TMI-2   490, 506, 510, 534, 634
Remote controlled video surveys, TMI-2   481, 614, 621, 631, 640
Restart, TMI-1   2, 51, 91, 93, 143, 149, 150, 158, 160, 173, 174, 192, 197, 212, 213, 222, 223, 230, 234, 236, 237, 290, 295, 300, 306, 313, 317, 321, 325, 327, 333, 336, 339, 469, 470, 583, 596, 597, 627, 652
Rogovin Report   120, 141

Severe Damage Analysis Package Computer Code, TMI-2   606
Sholly vs. NRC   165
Socio-economic effects, TMI-2 accident   9, 136, 193
Spectography, ramen   520
Spectography, luminescence   520
Spectroscopy, laser   592, 608
Spent Fuel Pool Piping System, TMI-1   583
Strontium releases, TMI-2   519, 589
Submerged demineralizer system, TMI-2   155, 172, 436, 475, 476, 486, 512, 524, 537, 610, 635, 654, 662, 665
Susquehanna River, TMI-2 accident, effect on   49

Technical Integration Office, TMI-2   438
Technology, transfer of   620
Telephone survey, TMI area residents   84
Tellurium releases, TMI-2   526, 557, 668
TMI Abnormal Waste Project Plan   527, 528
TMI Advisory Panel   88

TMI Dry Defueling System 558
TMI Instrumentation and Electrical Program 600
TMI Program Office 86, 87, 91
Title listing, by docket, TMI-1 documents 81
Title listing, by docket, TMI-2 documents 31, 78, 79, 80, 342, 343, 345
Transient Reactor Analysis Code (TRAC), TMI-2 573, 577, 629

U.S. Atomic Energy Commission see Atomic Energy Commission (U.S.)
U.S. Department of Energy see Department of Energy (U.S.)
U.S. Environmental Protection Agency see Environmental Protection Agency (U.S.)
U.S. Nuclear Regulatory Commission see Nuclear Regulatory Commission (U.S.)
U.S. Nuclear Regulatory Commission. Special Inquiry Group Report see Rogovin Report
U.S. President's Commission on the Accident at Three Mile Island (TMI) see President's Commission on the Accident at Three Mile Island (TMI)
U.S. Senate, Committee on Environmental and Public Works, Subcommittee on Nuclear Regulation, Report see Hart Committee report
U.S. National Weather Service see National Weather Service (U.S.)
Uranium recovery, TMI-2 541, 643

Western Psychiatric Institute and Clinic 11

York, County (PA) 28
York County (PA), radiological emergency response plan 28

Zeolite, Ionsiv A-51 625
Zeolite, Linde Ionsiv IE-95 586
Zeolite, Linde Ionsiv IE-96 625
Zeolite Vitrification Demonstration Program, TMI-2 440, 449, 450, 512, 525, 543, 586, 588, 610, 625, 641
Zirconium oxide 520, 626

THE LIBRARY
ST. MARY'S COLLEGE OF MARYLAND
ST. MARY'S CITY, MARYLAND 20686